T0340227

Freight Transport and Distribution

Freight Transport and Distribution

Concepts and Optimisation Models

Tolga Bektaş

CRC Press
Taylor & Francis Group
Boca Raton London New York

CRC Press is an imprint of the
Taylor & Francis Group, an **informa** business

CRC Press
Taylor & Francis Group
6000 Broken Sound Parkway NW, Suite 300
Boca Raton, FL 33487-2742

First issued in paperback 2019

© 2017 by Taylor & Francis Group, LLC
CRC Press is an imprint of Taylor & Francis Group, an Informa business

No claim to original U.S. Government works

ISBN-13: 978-0-4822-5870-7 (hbk)
ISBN-13: 978-0-367-87087-4 (pbk)

Visit the Taylor & Francis Web site at
http://www.taylorandfrancis.com

and the CRC Press Web site at
http://www.crcpress.com

This book is dedicated to my parents, Necla and
Yusuf Bektaş

Contents

Foreword

In western economies, freight transport accounts for about 10% of the gross domestic product and for an even larger portion of employment. However, these proportions have decreased in recent years because of a shift towards lighter goods, such as electronic products, and also because the transport sector is becoming more efficient. What these statistics do not tell is that without freight transport our economic system would be unsustainable, and civilisation, as we know it, would simply not exist.

The field of transport is truly multidisciplinary, as witnessed by the existence of university programs and scientific journals in areas as diverse as operational research, economics, engineering, computer science, regional science and geography. Quite significantly, transport planning has been a central area of investigation in operational research for more than half a century. Given the highly competitive environment in which firms now operate, it has become essential to optimise the delivery of transport activities through the use of sophisticated models and algorithms. The coverage provided by this book is therefore highly topical. The success of operational research in the transport area can be explained by several factors. First, mathematical programming offers a powerful paradigm for capturing the complexity and intricacies of transport problems. Second, the strong structure of these problems makes them amenable to solution by algorithms such as network optimisation and heuristic search. Third, the economic importance of transport problems is such that the gains yielded by the use of optimisation typically translate into increased efficiency and substantial monetary savings.

Readers will find in this book an introduction to the basic concepts of logistics, distribution and transport activities, as well as a comprehensive coverage of the main problems arising in freight distribution, such as location analysis, network design, classical vehicle routing, and problems that integrate related dimensions like inventory management and production. The book also describes the classical deterministic, stochastic and dynamic solution techniques applicable to freight transport.

Traditionally, transport studies have focused on cost minimisation, without much concern for the considerable amount of polluting emissions generated by the movements of goods. However, this is rapidly changing with the rise of public and political awareness for the environment. In recent years, researchers have put forward new models applicable to the reduction of pollution in freight transport. This line of research is rooted in the development of speed optimisation models in maritime transport and in the pollution-routing model for road freight distribution. This field of study is fast growing and is well reflected in the book. Finally, cooperation in supply chain management is increasingly being recognised as a means of reducing costs and of improving customer service. The book devotes a full chapter to this important topic.

This book is truly modern in its design, coverage and treatment. It can be used as a basic reference in a postgraduate course and should be of high benefit to researchers engaged in the field of freight transport.

Gilbert Laporte
Montreal, Quebec, Canada

Preface

A significant amount of progress has been made on research into planning problems for freight transport and distribution. The contributions made on the topic range from simple algorithms to elaborate mathematical models and sophisticated optimisation techniques. I may be so bold as to say that some of the problems in this area, such as the Travelling Salesman Problem or the Vehicle Routing Problem, have reached a state of maturity in terms of being well understood, formally defined and solved to a satisfactory standard. The field itself, however, is continually evolving, whether it be due to the advent of new information and communication architectures, novel technologies and unconventional means of transportation infrastructure, to address the new challenges ahead posed by social, economic and environmental considerations. The rapid changes bring with them new problems, some of which have already received attention from the research community, but also others that are yet to be properly understood and defined.

I may also go as far as saying that not all of the existing knowledge on this topic remains easily accessible to a wide range of audience, including undergraduate and postgraduate taught (master) students, as well as consultants and practitioners working in the area. One reason for this well-acknowledged divide is the fact that much of the research done in the area is at, and often well beyond, a postgraduate level of study. Much elaborate, innovative and technically sound they may be a high-level mathematical understanding and sophisticated coding skills required for implementation may limit the applicability of some of these methods. In contrast, mathematical programming remains a general, highly flexible and fairly intuitive way of formulating the wide range of planning problems in freight transportation and distribution. The advances in theory and practice of optimisation have made it possible to solve more intricate formulations easier than it was possible in the past, particularly in the light of the availability of a wide range of off-the-shelf optimisation software. Notwithstanding its limitations, mathematical programming still remains a useful tool, not only to conceptualise and formalise an optimisation problem but also as a starting point for solving these problems, at least partially or approximately, if not optimally.

This is precisely the line of thinking that I have taken in writing this textbook, the premise of which is the use of mathematical modelling to formulate (and ultimately solve) a number of optimisation problems arising in freight transport and distribution. This may explain why the textbook is structured as a collection of mathematical models for some of the fundamental planning problems in this area. It is again for the same reason that the book does not contain descriptions of bespoke algorithms, or techniques, for the optimisation problems covered here. The only exception is made in the methodology section, which presents optimisation algorithms of a generic nature that can be used to solve large-scale formulations, for the interested reader to pursue further.

It is by no means my intention to claim that this book fills the divide that has been referred to earlier, but I hope to be able to have made at least a small contribution towards bridging the gap between the theory and practice in the area. Indeed, a number of excellent textbooks have already been written in this area, some of which are referenced in this book. In this book, I hope to have covered some of the areas that perhaps may not have been given the same level of detail elsewhere, or areas for which the reader will simply find it useful to have an alternative coverage. To further contribute to applying theory to practice, I have provided numerical applications for most of the basic formulations presented in the book, as well as on the use of various types of constraints that appear in a number of formulations.

The book requires some basic knowledge on mathematical modelling, in particular linear and integer programming. The reader may also find it useful to familiarise themselves with at least one off-the-shelf optimisation software, in order that they are able to implement some of the formulations described in the book.

This book would not have been possible without the rewarding experience of my previous collaborations with colleagues across the globe, a list of whom would be too long to provide here. I have, wherever appropriate, provided references to joint work with my co-authors. I must, however, extend special thanks to Gilbert Laporte, not only for his valuable comments and suggestions on the first draft of this book but also for all the inspiration he has given me over the years. I thank Çağrı Koç, Dimitris Paraskevopoulos and Yi Qu for their input into some of the examples contained within the book, and Tri-Dung Nguyen for his comments on Chapter 7. Finally, I thank the publishers for their support, in particular Tony Moore, for his encouraging comments along the way.

Being a first edition, this book will no doubt contain inadvertent errors and unintended omissions despite all the care that has been taken, the entirety of which lies with the responsibility of its author. I hope to be able to rectify such errors or omissions in further editions.

Tolga Bektaş
University of Southampton
Southampton, England

A note on notation

Much effort has been made to standardise the notation used throughout the book as much as it practically proved possible. As a general convention, all decision variables and parameters have generally been denoted by lowercase letters (e.g. x_{ij}, π_k) of the Greek or the Latin alphabet, albeit minor exceptions apply. In contrast, sets and functions have generally been denoted by using uppercase letters. This convention necessitated breaking away from the somewhat conventional notation that is commonly used for particular problems, such as the uppercase letter Q, for example, that is often used to denote vehicle capacity in vehicle routing problems.

Parameters of the same or a similar nature that appear in different contexts or problems have been denoted using the same letter, so as to ensure some level of notational consistency. For example, the amount of demand requested by each customer node $i \in V_b$ in the facility location problem is shown by q_i, much in the same way as the demand of each customer $i \in V_C$ has been denoted by q_i in the vehicle routing problem. Node costs are shown by g, whereas link costs are denoted by c, with different indices used depending on the context. As for decision variables, x is generally related to the use of links and y for nodes (e.g., location). Finally, V is always used to denote a set of nodes (but with different indices such as V_d for demand points in the hub location problem and V_C for the customer nodes in the vehicle routing problem), P is the set of commodities and K is the set of vehicles.

The last chapter on methodology is an exception, in particular the notation used to describe the various algorithms therein should be treated separately from the rest of the book.

It is hoped that the reader will forgive any use of notation that causes any conflict across the chapters, which, in some cases, has been inevitable. In this case, the notation introduced in each chapter should be confined to the use within that chapter, and in isolation from the rest of the book.

Chapter 1

Freight logistics, distribution and transport

Concepts

The title of this chapter includes three terms, namely *logistics*, *distribution* and *transport*, all of which have a central role in the treatment of the topic, and as will be for the rest of this book. However, the precise meanings of these terms continue to cause some confusion. Various definitions have been put forward in the past which differ considerably. This is, however, understandable not only because their meaning tends to change depending on the industry in which they are applied, or even evolve over time, but also due to the overlap in the elements they cover. In the following, we will attempt to differentiate these terms, not for the purpose of arriving at yet another set of definitions, but to reduce any element of confusion for the reader of the book, at least in the way that they will be used in the remainder of the exposition:

- *Transport* is the actual movement of goods from one location to another using a means or a vehicle of transport (e.g. trains, trucks, boats) and a transport infrastructure (e.g. roads, railways, canals).
- *Distribution* often denotes all activities relevant to physical movement of goods, including transportation, but also transhipment and warehousing.
- *Logistics* is generally used as an overarching term that includes all activities related to the movement and coordination of goods from their source of origin to the final point of delivery, and includes production and distribution. Here, movement does not just correspond to physical movement of goods but also the flow of information.

1.1 Actors

While freight transport and distribution include many elements, including various organisations and people, the three fundamental actors who actively take part in the domain are described here.

1.1.1 Shippers

The demand for freight transportation is generated by shippers. Each shipper will have their own logistics strategy, which includes whether to operate their own fleet or to use an external party to which it will outsource their logistics and distribution activities, as well as choosing the mode(s) of transport. The process through which the shippers will define their logistics activities generally takes a three-level decision structure:

1. The *long*-term decisions in the first level involve defining strategies in line with their customer network and production activities.
2. The second-level *medium*-term decisions include levels of inventories at production, warehousing, and distribution facilities, frequency and amount of shipping and flexibility of service.
3. At the *short*-term level, shippers decide on the attributes of the services required for its shipments, such as maximum rates, transport time, reliability and safety.

In making these decisions, they will consider the availability and the characteristics of the services offered on the market by carriers and intermediaries, such as freight brokers and third-party logistics providers.

1.1.2 Carriers

Carriers are people, businesses or organisations that operate and offer transportation services for shippers. They may either provide a customised service, where a vehicle or a fleet will be dedicated exclusively to a particular customer, or operate on the basis of consolidation, where each vehicle contains several pieces of freight for different customers with possibly different origins and destinations. In the latter case, carriers generally operate their services according to a published timetable, which prescribes routes, schedules and rates they offer.

1.1.3 Intermediaries

In some cases, the shipper operates their own fleets of vehicles and does not require an external carrier to ship goods on their behalf. In this case, the management of the relevant transportation and distribution is done in-house. If a shipper does not own a fleet, then it may choose to work directly with one or several carriers.

Shippers may alternatively use a *freight forwarder*, an intermediary person or organisation that acts as a third party and manages the shipments on behalf of the shipper by contracting one or several carriers. They also help to identify a suitable mode or a combination of modes for the shipper. Freight forwarders work closely with shippers and carriers, as well as other entities in the transportation network, such as ports or terminals, particularly if they additionally undertake ancillary services such as customs clearance.

1.2 Modes of transportation

There exist different means of transporting freight over the network, each of which is referred to as a *mode* of transport. Transportation modes can be differentiated with respect to the type and specification of the vehicle used, the underlying technology, the relevant infrastructure and the nature of the associated operations. The three main modes of transport are air (e.g., cargo planes), land (including road, rail and off-road) and water (e.g., ships in oceans, barges in rivers). Other modes of transport, such as pipelines (e.g., to transport gas) and cable transport (e.g., elevators and cable-cars), also exist.

The term *mode* can also be used to denote different types of vehicles within a given domain of transport. For example, trucks, vans and bicycles can be seen as three separate modes operating within road transportation due to their distinct features, such as different capacities, capabilities and restrictions. In this example, trucks have the largest capacity but often have restrictions in travelling in urban areas, whereas bicycles are much smaller in capacity but do not suffer from the same type of restrictions as trucks or vans. Brief descriptions of the three main modes of transportation are explained here.

1.2.1 Road

Road has been, and continues to be, *the* most widely used mode of freight transport, both nationally and globally. One of the main reasons behind its popularity is the ability of road transport to offer a very quick service and often be available on demand.

A wide variety of vehicles are used for road freight transportation, which can be differentiated on the basis of size, capacity, weight and the type of energy used. Vehicle classification on road transportation is generally based on the *Gross Vehicle Weight Rating (GVWR)*, which refers to the maximum allowable total weight of a vehicle including its empty mass, fuel and any load carried. The empty mass of the vehicle, but with fuel and fluids such as engine oil, is named as the *curb weight*. Vehicle classifications vary from one country to another. In the United States, eight classes exist, with vehicles in the lightest class having a GVWR up to around 3 tonnes, and those in the heaviest class with a GVWR higher than 15 tonnes. In the United Kingdom, more classes exist, with those of at most 3.5 tonnes gross weight described as light goods vehicles (LGVs) and those between 3.5 and 44 tonnes gross weight named as lorries or heavy goods vehicles (HGVs).

Most vehicles used in road freight transportation run on gasoline or diesel engines. Vehicles using alternative sources of fuel or energy have also been developed, such as those running on batteries, biofuels (such as bioalcohol or ethanol), biodiesel, compressed natural gas, hydrogen, and liquefied petroleum gas (LPG), for use in freight distribution. Within urban areas, human-powered vehicles, such as bicycles and tricycles, can also be used for goods deliveries. To overcome the sole dependency on human power, some of the freight bicycles have power assist motors to aid the cyclist.

The road network is composed of motorways (or highways), urban roads, rural roads, lanes or graded roads and includes bridges and tunnels. Traffic on the road network is controlled by means of traffic signals, signs or markings on the pavement. Various legal requirements are imposed on freight vehicles travelling on the road network, which include limitations on vehicle weight, dimensions, mandatory equipment, licences and insurances (Rushton et al., 2014). As for truck drivers, there also exist regulations on driving and working hours, which restrict the duration of driving time and require break and rest periods in long-haul journeys. These regulations aim at reducing driver fatigue, which is known to have adverse affects on road and driver safety. The regulations usually differentiate between *on-duty* time, which is the time spent working, including driving, waiting, loading and unloading and doing paperwork, and *off-duty* time, where the driver has no obligation to work. In the United States, for example, these regulations are known as *Hours of Service*, which limit the maximum consecutive driving time between two rest periods to 11 hours, at which point the driver must be off-duty for at least 10 consecutive hours. Furthermore, a truck driver cannot drive if 8 hours or more have elapsed since the end of the last off-duty period of at least 30 minutes. Similar regulations prevail in other countries, albeit with differences (Goel and Vidal, 2013).

1.2.2 Rail

Rail freight transportation is known for its ability to offer cost-effective long-haul transportation services, primarily, but not exclusively, for bulk cargo. There are two major components of a rail system, namely the rail network infrastructure and freight trains.

The rail network is a large and complex structure composed of *nodes* and *tracks* (or track segments) as links between the nodes. The former include *yards* or *terminals* where classification or marshalling operations are performed, *stations* where cargo is picked up from or delivered to and *junctions* that are signal-controlled points in the rail network to allow trains to switch from one route to another.

Freight trains are composed of one or more *locomotives*, and several rail *wagons* (or *cars*). Locomotives move the train along the tracks by either pulling it from the front or push from the rear and range from the earlier types powered by steam to contemporary ones using electricity, magnetic force or diesel engine. Rail wagons carry the freight and come in a variety of forms, including specialised wagons for carrying particular types of cargo (e.g., autoracks for carrying automobiles or refrigerator cars for temperature-sensitive goods). A train is characterised by its route, origin, destination, intermediate stops, the physical path it travels on and the schedule information that includes departure and arrival times at each station where it stops. Each wagon also has an itinerary that specifies an origin and a destination station and need not correspond to the origin and destination of the train on which it is carried. Wagons may travel on several trains during their journey, usually in groups called *blocks*. Each block is assigned an origin and a destination, although individual cars in a single block may have different origins and destinations. A block is treated as a single unit for handling purposes. Once formed at its origin yard, a block will not be classified again until it arrives at its destination yard.

Classification or *marshalling* refers to a set of operations carried out at yards or terminals, where incoming trains are disassembled by decoupling the rail cars and new trains are formed using individual cars or blocks. Bektaş et al. (2009) provide a detailed description of the operations at a classification yard, according to which a train arriving at a yard first enters a receiving area, where the engines are taken away for inspection and maintenance, blocks are separated and cars are inspected. The classification operation begins from this point on and can be performed in two ways, depending on the type of the rail yard. In *flat* yards, a switching engine is used to push a group of cars out of the receiving tracks onto one of the classification tracks. In *hump* yards, classification is performed by using an artificially built hill, called the hump, where an engine pushes a group of cars out of the receiving area and up the ramp until it reaches the top of the hill. Due to the pull of the gravitational force, the cars roll down the incline on the other side of the hump, usually one car at a time, and are directed onto one of the classification tracks. Following this operation, each classification track becomes occupied by a group of cars that form the block. Each block then waits until the departure time of its outbound train. When the train is due to leave, they are pulled out of the classification tracks onto the departure tracks and are attached to the train. Following one last inspection of the whole train, the train and blocks leave the yard. Figure 1.1 shows the general structure of a classification hump yard.

Classification is not the only operation that is performed at a rail yard. Other types of operations include inspection, crew change, refuelling the trains and dropping and picking up blocks of cars. Among these, however, classification is known to be the major time-consuming operation.

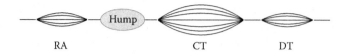

Figure 1.1 A schematic representation of a rail classification (hump) yard showing the receiving area (RA), classification tracks (CT) and departure tracks (DT).

1.2.3 Air

Air transportation is mainly used for goods that are time sensitive or have high value-to-weight ratios, including perishables due to the need for speed, fashion goods, emergency supplies and spare parts (Rushton et al., 2014). Speed is a major advantage of air transportation but comes with the significant drawback of having a very high cost. This is one of the reasons as to why air transport has always had a marginal share in the amount of freight shipped worldwide, in comparison to other modes of transport.

Air freight transportation uses a system of air hubs, where outgoing cargo to be shipped is received and consolidated for shipping to a different hub in the network or incoming cargo from another hub is separated and shipped to its final destination. The network tends to have a hub-and-spoke structure (see Chapter 2) to make the best use of economies of scale and help to achieve the lowest possible cost of transport. Specialised pallets or containers named *unit load devices* are used to stow cargo in aircraft, in a safe and efficient way. Air cargo can be carried in passenger planes along with passenger baggage. Alternatively, it can be transported in cargo aircraft designed specifically for freight, or by helicopters for access to areas that are difficult to reach.

1.2.4 Sea

Sea or maritime transport is the dominant mode with which a large majority of international trade is moved. It is suitable for transporting bulk goods, large packages or containers and similar types of cargo that are not time sensitive.

Vessels are the main vehicles of maritime transport, ranging from small boats to larger ships. Some of the common ship types include *break-bulk vessels* that are used to transport any type of loose cargo excluding liquid or loose bulk commodities (e.g. boxes, barrels, casks and bagged cargo), *Roll-on/Roll-off* (*RoRo*) vessels that carry wheeled cargo (e.g. cars, trucks, trailers), *oil tankers* that carry crude oil in liquid form, *dry bulk vessels* that carry bulk goods (e.g. coal, ore, grain) and *reefer ships* that carry perishable goods that require refrigeration (e.g. fish, meat, fresh produce).

The maritime transportation network consists of *ports*, which are hubs where ships depart from, stop at, or return to, used for loading, unloading and handling cargo. Ports come in different types based on their location and the nature of the cargo they handle. Seaports are the most common. Some ports only work with containerised cargo and include specialised equipment such as a *gantry crane* used to transfer containers between a vessel and other vehicles of land-based transport, such as trucks and rail wagons, and a *stacker* that is used to lift and swap containers, as well as perform loading and unloading operations on trucks and wagons.

In maritime transportation, there are larger variations in travel as compared with other modes of transport, caused by, for example, weather conditions or congestion at ports. In addition, travel and loading/unloading times are longer, particularly in the case of intercontinental trade.

There are three main modes of operation in maritime transportation, as categorised by Christiansen and Fagerholt (2014):

1. *Liner shipping* is where vessels operate on regular routes with fixed schedules. The routes and schedules are published in a timetable, similar to bus services for passenger transportation. The routes are fixed, each of which is called a service route. There may be several vessels operating on a given service route, and this depends on the desired frequency over a given period of time (e.g., weekly).
2. *Industrial shipping* is where the cargo owner operates their own vessel fleet.
3. *Tramp shipping* is where a carrier operates a fleet of vessels to transport *mandatory cargo* between specified ports and within a certain time frame for contracted shippers but can also optionally carry spot cargo for other shippers, in order to increase their revenues. Tramp operators will need to decide on whether to accept or reject any spot cargos, which will depend on capacity and timing considerations in the light of the already scheduled routes for mandatory cargo.

Agarwal and Ergun (2008) draw a useful analogy between liner shipping and a passenger bus service, between industrial shipping and personal cars and between tramp shipping and a taxi service.

1.2.5 Intermodal transportation

A transportation network is said to be of *unimodal* type if it only operates a single mode of transport or of *multimodal* type if it operates at least two different modes of transport. Unimodal transport cannot take advantage of the use of several modes of transportation, and it is either impractical or impossible to assume a single mode of transport for performing door-to-door delivery. If freight originates from an inland warehouse, for example, and is destined to another inland warehouse on another continent, then part of the journey will have to involve air or water transport.

However, even if there are no physical or geographical limitations, carrying small amounts of cargo on a relatively large vehicle from origin to destination is inefficient. One way to improve efficiency is consolidation. The aim of *intermodal transportation* is exactly this, that is, to consolidate loads for efficient long-haul transportation (e.g., by rail or large ocean vessels), while taking advantage of the efficiency of local pick-up and delivery operations by truck. For any shipment to be intermodal, freight must be transported from origin to destination by a sequence of at least two modes of transport, where transfers from one mode to another take place at *intermodal terminals* (such as sea ports or rail yards). An illustrative example is provided here.

Example 1.1

A multimodal transportation network is illustrated in Figure 1.2, on which three modes of transportation, namely road (trucks), rail and water (shipping), operate. Cargo originates from the two facilities seen on the far left of the figure and is destined to the warehouses appearing on the far right of the figure. The figure also shows how the cargo is shipped using an intermodal chain of modes. In particular, loaded trucks leave the origin facilities to a rail yard, where they are consolidated into a train and sent to another rail yard. Trucks are again used to transport the containers from this rail yard to a sea container terminal. This last operation may not be necessary if the sea container terminal has an interface to the rail network, in which case freight is transferred directly from the former onto the latter. Containers are then transported to a port on another continent by ocean shipping, from where they leave by either trucking or rail (or both) to their destinations.

One of the advantages of intermodal freight transport is the possibility of *modal shift*, which is defined as partially or fully transferring freight from one mode to the other in the network when it is being shipped to its final destination.

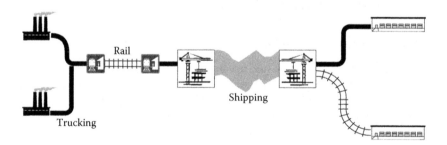

Figure 1.2 An example of an intermodal transportation network.

1.2.6 Intermodal terminals

In intermodal transportation, transfers from one mode to another, and all associated operations related to, for example, loading and unloading, temporary storage or intermediate buffer, and even pre-delivery inspection or enhancement work on the goods are carried out at *intermodal terminals*. Here are some examples of intermodal terminals:

- *Rail yards* where transfers between the two modes of road (e.g. trucks and lorries) and rail take place.
- *Sea ports* that act as the interface between sea transport (vessels, ferries) and land transport (road or rail). A barge terminal is a smaller port used in inland river transportation and is linked to larger sea terminals.
- *Inland distribution centres* that are used only for road transportation, and act as transfer points between different modes of road transport, for example lorries and vans. Warehouses and cross-docking centres are examples to inland distribution centres.

Terminals are perhaps the most critical components of an intermodal transportation chain, as the efficiency of the latter highly depends on the speed and reliability of the operations performed in the former. Avoiding unplanned delays and the formation of load or vehicle bottlenecks is one of the major goals in operating intermodal terminals.

1.2.6.1 Containerised transport

Cargo flow in an intermodal transportation network can be carried by a variety of means, for example crates, pallets or boxes for fresh produce, specialised vessels or tanks for liquid bulk cargo. Irregularity of general cargo, in shape or size, has led to the development of standardised units of shipment, which came to be known as containers. A container is a steel or aluminium box, with a standard size of 20 ft in length, known as the Twenty Foot Equivalent Unit (TEU), is widely used around the globe, but 40 and 45 ft containers are more commonly used in North America. Containers offer a number of advantages over non-containerised cargo, which are summarised as follows:

1. Containers can be locked, which means that their contents cannot easily be modified except at origin or destination, which implies increased safety and reduced loss and damage.
2. Due to its standard structure, transfer operations at terminals can also be tailored to process containerised cargo fast and with minimal amount of effort. Container ports are excellent examples where such operations

can be seen, using container-specific equipment such as container gantry cranes, implying reduced cargo handling.
3. Containers can be stacked, which implies a more efficient use of storage space within the port.
4. Cargo of various sizes, shapes, dimensions or weight can be carried in a container, making it a flexible enough means of transport to accommodate for a variety of cargo types.

Due to these advantages, intermodal transportation, as it is implemented and used today, heavily relies on containerisation, to the point that intermodal transportation is often equated to moving containers over long distances, on multimodal networks. Intermodal transportation is not restricted, however, to containers and intercontinental exchanges.

1.3 Choice of carrier and transportation mode

There are a number of factors that shippers use to decide which particular mode or modes of transportation to use for shipping their freight. Earlier studies on this topic have identified six main factors that influence a shipper's decision:

1. *Freight rates* (including cost and charges)
2. *Reliability* (delivery time)
3. *Transit times* (time-in-transit, speed, delivery time)
4. *Over, short and damaged shipments* (loss, damage, claims processing, tracing)
5. *Shipper market considerations* (customer service, user satisfaction, market competitiveness, market influences)
6. *Carrier considerations* (availability, capability, reputation, special equipment)

Later research identified *frequency* and *flexibility* as additional factors in particular industries, and that factors such as *international dimension, economies of scale, security, environmental concerns and energy use, integration with the supply chain* and *information technologies*, and particularly the use of the *Internet* are all relevant to the transportation mode choice but remain as under-researched areas.

Other studies have recognised that the perception of a mode of transport is likely to have a significant effect on the choice of mode. Of the more prominent ones, earlier research has identified *communication, quality of service, consistency* (in delivery), *transit times* and *rates* as being the main factors, with the first two being more significant than the others.

1.4 Shipment options

There are at least two ways in which shipments can be made, which are described in more detail here.

1.4.1 Direct and customised shipments

Direct or customised shipments arise when goods are shipped from the source of supply to the demand point without the use of any intermediate facilities or demand points. Full-load trucking is an example of customised transportation, where, upon the call of a customer wishing to have goods shipped from one (source) point to another (destination) point, a dispatcher assigns a truck to this task. The truck travels to the customer location, is loaded and then moves to the destination, where it is unloaded. Following this, the driver is assigned a new task by the dispatcher, kept waiting until a new demand appears in the near future, or repositioned to a location where a load exists or is expected to be available about the arrival time. The advantages of full-load trucking come from its flexibility in adapting to a highly dynamic environment and uncertain future demands, offering reliability in service and low tariffs compared to other modes of transportation. The fill efficiency of full-load trucking is achieved through the implementation of resource management and allocation strategies that seek to make the best use of the available resources, while maximising the volume of demand satisfied and the associated profits. Customised services are also offered, for example, by charted sea or river vessels and planes.

1.4.2 Consolidated shipments

In many cases, trade-offs between volume and frequency of shipping, along with the cost of transportation, render customised services impractical. In such cases, consolidation is used for serving a number of demand points using a particular service. Freight consolidation transportation is performed by Less-Than-Truckload (LTL) motor carriers, railways, ocean shipping lines, regular and express postal services, etc. One example of LTL transportation is when the loads of a given set of customers are consolidated at a given source (e.g. a depot), and the deliveries are multiple destination points. In this case, a truck or a lorry is loaded with all the goods at the depot and is dispatched to visit the customers in a particular order to perform the deliveries. A consolidation-based transportation system can also be structured as a hub-and-spoke network, where shipments for a number of origin-destination points may be transferred via intermediate consolidation facilities, or hubs, such as airports, seaport container terminals, rail yards, truck break-bulk terminals and intermodal platforms.

1.5 Distribution structures

The structure of the way in which goods are distributed is generally in the form of single-echelon (or single-tier) or multi-echelon (or multi-tier) distribution network. The type of structure used depends on a number of factors, such as the type of area (e.g. urban, suburban, rural) or the size of the area (e.g. countries, continents), the type of product(s) being shipped, the types of vehicles used and the demand requirements in terms of volume and time.

1.5.1 Single-echelon

A single-echelon distribution structure does not involve the use of any intermediate facilities between the source(s) of supply and the source(s) of demand and operates on the basis that deliveries are made from the former to the latter. This might be in the form of direct shipments from the supply point to the customer, or, as in the case of consolidation, from one supply point to many customers.

Figure 1.3 shows a single echelon distribution structure where a central depot, shown by D, serves five customers labelled C_1–C_5, all using direct shipments. In this example, it is assumed that each link carries at least one full-truckload shipment. Figure 1.4 shows the same network in which it is assumed that the deliveries can be consolidated. In particular, one LTL vehicle serves customers C_1 and C_4 in the given order, and the other serves customers C_3, C_5 and C_2 in the given order.

1.5.2 Multi-echelon

If the distribution network includes several layers of intermediate facilities, such as warehouses, consolidation facilities, distribution centres and

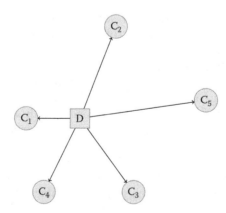

Figure 1.3 A single-echelon distribution network structure with direct shipments.

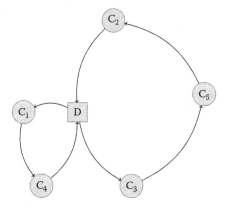

Figure 1.4 A single-echelon distribution network structure with consolidated shipments.

cross-docks, to which the goods are sent as part of the distribution from their origin(s) to their destination(s), then the network is said to have a multi-echelon structure. Supply chain networks often have such a structure, for which the planning problems entail the management and operation of the facilities in the network as well as the assignment of (aggregated) flow between the facilities. In the context of freight distribution, however, the planning problems are more specific to the management of transport modes and the vehicle fleets used in between the different layers.

A two-echelon network distribution structure is shown in Figure 1.5 where the first echelon includes all movements between the first layer of depots shown by D_1 and D_2 and the second layer of intermediate points (e.g. warehouses) shown by W_1, W_2 and W_3. In this example, consolidated

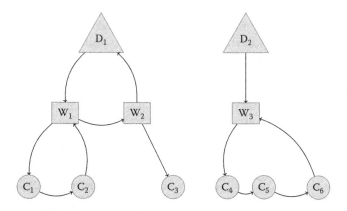

Figure 1.5 A two-echelon distribution network structure with mixed types of shipments (direct and consolidated).

shipments depart from D_1 to serve the two warehouses W_1 and W_2. In contrast, a direct shipment using a full-truckload is made between D_2 and W_3. The second echelon consists of all the distribution activities between the second layer and the third layer, where the latter includes six demand points shown by C_1–C_6. It is clear from the figure that this network uses a mix of consolidated and direct shipments at this echelon.

Structures such as those shown in Figure 1.5 have been suggested for use within urban areas, where the first layer of depots consists of those that are located outside the boundaries of the urban zone and the second layer consists of intermediate points located on the boundaries. The expectation is that bulk deliveries are made from the first to the second layer using heavy goods vehicles, which would stop here. The distribution activities within the urban zone would then be carried out using smaller vehicles that would operate between the second layer of intermediate points and the third layer of demand points. Such a strategy prevents large and heavy vehicles from entering urban zones.

References and further reading

Transport, logistics and distribution fall within the wider remit of supply chain management, a topic which is beyond the scope of this book. Interested readers are referred to Simchi-Levi et al. (2009) for an introduction to supply chains. A detailed and comprehensive treatment of logistics and distribution management, including concepts, planning, procurement and inventory decisions, warehousing and storage and operational management is given in Rushton et al. (2014), which also presents a good coverage of the practical aspects of various modes used in freight transport. For an excellent coverage of logistics systems management, including use of forecasting methods, locating and managing facilities, supplier selection and freight transport management, and the relevant mathematical models and solution algorithms, see the textbook by Ghiani et al. (2013). A detailed treatment of various problems arising in transport and distribution, ranging from public transit and passenger railway optimisation to traffic equilibrium and hazardous materials' transportation, can be found in the edited volume by Barnhart and Laporte (2007), in particular see Cordeau et al. (2007) for road transport and Christiansen et al. (2007) for maritime transport. An introduction to intermodal transportation is given by Bektaş and Crainic (2007) and Crainic (2007). For studies on mode choice and factors affecting the decisions, see McGinnis (1990) and Murphy and Hall (1995). Shipper perceptions are discussed by Evers et al. (1996) and Evers and Johnson (2000). For a review of this area of research, see Waller et al. (2008).

Chapter 2

Location in networks

Large-scale transportation networks are often characterised by the amount of freight that needs to be shipped between various points on the network, and by the resources available within the existing infrastructure on which shipments are to be made. Strategic decisions that concern the investments made into the design of a new network or the enhancement of an existing network have long-lasting effects, spanning years or even decades. The nature and amount of the investments made, in turn, depend on the demand generated by the users on the network. Here, demand corresponds to any type of good, such as parcels, postal mail or express packages, and is generally associated with a pair of origin and destination nodes, and with a given size (e.g. volume, mass or density). The transportation of such goods from one point to another in the network is performed by vehicles, such as trucks, trains, vessels or airplanes.

Given the limited amount of resources available, it is often difficult, or expensive, to commit individual vehicles to serve a particular origin–destination (OD) pair to provide direct shipments. Customised transportation services may be suitable for environments in which the parameters, including shipment size, dispatch times and delivery dates, are subject to high variability, and where there is uncertainty in future requests. These issues are relevant to operational planning and are difficult to account for within the context of strategic planning which assume that the parameters are relatively stable over a long time period.

A more efficient way to plan shipments with different origins and destinations, particularly for when part of the shipment is to be done over long distances, is to use consolidation as explained in Chapter 1. For this purpose, investments can be made to establish *consolidation points*, or *hubs* on the network, where incoming flows into a hub will be consolidated into larger flows within the hub. The consolidated goods are then shipped to other hubs in the network using regular inter-hub services that can carry shipments of large sizes and often faster than services on links between hubs and non-hub nodes. Hubs can be airports, seaport container terminals, rail yards, truck break-bulk terminals and intermodal platforms. Due to the economies of scale, shipment between hubs are generally more cost-effective than direct shipments between non-hub points, implying a lower shipping cost per unit of flow. The activities taking place at the hubs include aggregating (consolidation), disaggregating (breakbulk) and redirecting (switching, sorting or connecting) flows. The resulting configuration is typically named as a *hub-and-spoke* network. In hub-and-spoke networks, low-volume demands are first moved from their origins to a hub where the traffic is sorted (classified) and grouped (consolidated). The aggregated traffic is moved in between hubs by high-frequency and high-capacity services. Upon arrival to the next hub, loads are transferred to their destination points from the hub by lower-frequency services, often using smaller vehicles. When the level of demand is sufficiently high, direct services may be run between a hub and a regional terminal. It should be noted that, although a hub-and-spoke network structure is likely to ensure a more efficient utilisation of resources and lower costs for shippers, it may also result in a higher amount of delays and a lower reliability due to the longer routes used and the additional operations performed at terminals.

Figure 2.1 shows a sample hub-and-spoke network with three hubs shown by H_1, H_2 and H_3, and with six origin or demand nodes denoted C_1–C_6.

In the example of Figure 2.1, nodes C_1 and C_2 are assigned to the hub shown by H_1, nodes C_3 and C_4 are assigned to the hub shown by H_2 and nodes C_5 and C_6 are assigned to the hub shown by H_3. For example, all flows leaving nodes C_1 and C_2 will be consolidated at H_1 for forwarding onto other hubs. Similarly, any flow destined to these two nodes will first have to flow into H_1, where it will be disaggregated and forwarded onto the node it is destined to via the links shown. In this example, all three hubs are interconnected but this does not necessarily have to be the case. Furthermore, it is sometimes possible to assign one point to more than one hub, which shall be discussed further in the chapter.

The combined decisions of where to locate a given number of hubs on a given transportation network and how to route the flow such that all

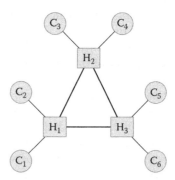

Figure 2.1 A typical hub-and-spoke network configuration.

shipments can be sent from their origins to destinations have given rise to a rich field of research known as *hub location* problems. The problems have common characteristics but differ with respect to a number of criteria, including the way in which OD nodes are assigned to hubs, the objectives and the costs involved. This chapter will describe several fundamental problems that have been studied in the field of hub location and present optimisation models for each of the problems.

2.1 Hub location problems

There are several distinct features of hub location problems, which are discussed here:

- *Underlying network structure*: Hub location problems are defined either on a network or a plane. In the former problem, named as *discrete hub location*, there exists a given set of nodes, which make up the OD pairs, and a finite number of potential locations on which hubs can be located. The latter problem, generally referred to as *continuous hub location*, assumes that hub facilities can be located anywhere on a given plane.
- *Flow*: Every OD pair in the network is associated with a nonnegative amount of flow. Hubs can be allowed to be origins or destinations of certain flows, or they can be defined exclusively as transhipment points where flows pass through and are processed at, implying that no hub can be located on an origin or a destination node.
- *Shipment options*: Direct shipments between OD pairs are not allowed, implying that all shipments must be routed through hubs. Hubs are fully interconnected, and a given flow needs to be shipped, from its origin to its destination, using at least one hub.
- *Costs*: Every link in the network is associated with a unit shipping cost, which might be distance, time or the actual travel cost incurred. The economies of scale exploited by the use of hubs imply that the unit shipping cost between the hubs is less than the unit shipping cost between any OD node and a hub, often discounted by a rate α. Furthermore, if the network costs satisfy the triangle inequality (as will be explained later), then a given flow need not visit more than two hubs on its way from the origin to the destination.

The requirements of the specific context in which hub location problems arise give way to a number of basic problems, which differ according to the following criteria:

- *Allocation* or *assignment*: The *single-allocation* or the *single-assignment* hub location problem refers to the case where a flow originating from an origin or a destination node can only be sent to a single hub, and a destination node can be supplied from a single hub. In contrast, the *multiple-allocation* or the *multiple-assignment* breaks away from this restriction by allowing origin or destination nodes to be served by more than one hub.

- *Capacity*: Hubs are facilities at which goods are processed using the operations described earlier, and they will often have capacity limits as to the amount of flow that a hub can accommodate over a given period of time. In this case, the problem is said to be *capacitated*. Alternatively, if there are no capacity constraints or the amount of flow passing through any hub can never be greater than its capacity, then the problem is said to be *uncapacitated*.

- *Objectives*: Various objectives have been used for hub location problems, including the ones listed here:

  - While the main objective of some hub location problems has traditionally been to minimise a function of total cost, the way in which this function is defined can vary from one problem to another but usually includes one or more of the following:

    - Transportation cost expressed as function of flow and the unit shipping cost
    - Fixed installation cost of locating a hub at a potential hub node
    - Fixed installation cost of establishing a link between OD nodes and hub nodes

  The *min-sum* (or *minisum*) terminology adopted from facility location can be used to refer to such class of objective functions.

- In problems where hubs need to be located to cover the total demand in the network, and where demand nodes need to be located within a certain radius of the location of the hub, an objective function is used that minimises the number of hubs. This objective does not consider the fixed installation costs of the hubs located.

- In some applications involving emergency facilities or perishable items, it is desirable that the plans avoid prescribing shipments that are overly high in cost, for example, too lengthy in distance or too long in time. In most cases, this can be ensured by an objective that minimises the maximum cost incurred across all the shipments. The cost might be defined separately for any OD pair, or for a particular shipment sent either on an inter-hub link or on a link between a hub and a non-hub node. In the standard facility location terminology, these problems are said to be of the type *min-max* (or *minimax*).

- *Number of hubs*: Some hub location problems require that exactly p hubs should be located on the network, whereas there is no such restriction for other types of problems.

2.2 Common notation

All problems and models described in this section are defined on a graph $G = (V, E)$, where the node set $V = \{1, 2, \ldots, n\}$ includes all the origin and destination nodes, as well as other nodes on which hubs can be located. We use V_d to denote the former set and V_h to denote the latter set. Using this notation, we can describe at least two types of hub location problems, as shown here:

1. $V_h = V_d = V$, denoting the general case where hubs can be located at any node in the network
2. $V_h \cup V_d = V$ and $V_h \cap V_d = \emptyset$, where hubs cannot be located on origin or demand nodes

We further define E as the set of connections between nodes $i \in V$ and $j \in V \setminus \{i\}$, each associated with a positive unit shipping cost c_{ij}. Note that this definition assumes that the direction does not matter, that is, $c_{ij} = c_{ji}$ for all $i, j \in V$, $i \neq j$. However, if the costs c_{ij} and c_{ij} are different for a given link $\{i, j\}$ in the set E, or when the route of a flow is specified with respect to an explicit ordering of nodes, then the problem is defined on a directed graph $G = (V, A)$ where A is the set of arcs. A total of w_{ij} units must be shipped from an origin node $i \in V_d$ to a destination node $j \in V_d$. We also assume that a discount rate α is applied to inter-hub shipment costs, where $0 \leq \alpha < 1$.

2.3 p-Hub median problems

A fundamental problem in hub location is locating exactly p hubs under a min-sum objective, which minimises the overall shipping costs associated with the flows. In this case, the demand points are assigned to one or more hubs that have been located, and there are no capacity limitations on the amount of flow processed by a hub, that is hubs are uncapacitated. There are two variants of this problem, which are covered in the two sections that follow.

2.3.1 Single-allocation p-hub median problem

The first variant is called *single-allocation p-hub median problem* where a demand node can only be assigned to a single hub. We present two types of formulations for this problem.

2.3.1.1 Quadratic integer programming formulation

A natural way to formulate the problem is to use binary variables x_{ik}, defined for each $i \in V_d$ and $k \in V_h$, that is equal to 1 if node i is assigned to hub k. If $V_d = V_h$, then a variable $x_{kk} = 1$ indicates that a hub has been located on node k. If $V_d \cap V_h = \emptyset$, then an additional set of binary variables are used. In particular, a binary variable y_k is equal to 1 if node $k \in V_h$ is a hub node and equal to 0 otherwise. We will first present a formulation assuming the latter case. The formulation also assumes that any OD flow originating from node $i \in V_d$ and destined to node $j \in V_d \setminus \{i\}$ is first sent to a hub $k \in V_h$, from where it is shipped to hub $m \in V_h$ before it is forwarded to its final destination. The formulation is as follows:

$$\text{Minimise} \sum_{i,j \in V_d; i \neq j} w_{ij} \left(\sum_{k \in V_h} c_{ik} x_{ik} + \sum_{m \in V_h} c_{jm} x_{jm} \right) \tag{2.1}$$

$$+ \sum_{i,j \in V_d; i \neq j} w_{ij} \left(\sum_{k,m \in V_h} \alpha c_{km} x_{ik} x_{jm} \right) \tag{2.2}$$

subject to

$$\sum_{k \in V_h} y_k = p \tag{2.3}$$

$$\sum_{k \in V_h} x_{ik} = 1 \qquad \forall i \in V_d \tag{2.4}$$

$$x_{ik} \leq y_k \qquad \forall i \in V_d, k \in V_h \tag{2.5}$$

$$x_{ik} \in \{0, 1\} \qquad \forall (i, k) \in E \tag{2.6}$$

$$y_k \in \{0, 1\} \qquad \forall i, k \in V_h. \tag{2.7}$$

The objective function is represented by the terms (2.1) and (2.2). The former includes two terms, the first of which accounts for the cost of sending flows between node $i \in V_d$ and hub $k \in V_h$, and the second is the cost of shipment from hub $m \in V_h$ to node $j \in V_d$. In (2.2), the cost represented is that of the inter-hub transportation $k \in V_h$ to $m \in V_h$. It is possible that $k = m$.

The objective function is quadratic due to the term $x_{ik} x_{jm}$, which is equal to one if and only if the flow follows path $i - k - m - j$, in which case the inter-hub shipment cost αc_{km} is included in the overall calculation. Constraints (2.3) guarantee that exactly p hubs are located on the network. The single-assignment restriction is modelled through constraints (2.4). Constraints (2.5) ensure that an assignment to node $k \in V_h$ is made only if a hub has been located at this node. Finally, constraints (2.6) and (2.7) model the integrality restrictions on the assignment variables.

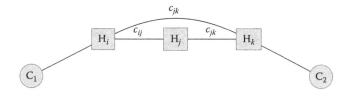

Figure 2.2 Two versus three hubs under the triangular inequality.

That the formulation only allows at most two hub visits for a given OD flow may be restrictive for any cost structure, as an optimal path might visit more hubs. However, if the costs satisfy the triangle inequality, namely $c_{ij} + c_{jk} \geq c_{ik}$ for any node triplet (i, j, k), then for any flow path using three or more hub nodes, one can find a path visiting at most two hub nodes whose cost will be no higher than the cost of the former. Such an example is shown in Figure 2.2, where the triangular inequality allows moving any flow on H_i–H_j–H_k onto H_i–H_k without the need to resort to using hub H_j.

An alternative and a more compact version of the earlier quadratic formulation can be written where constraints (2.5) are replaced by their aggregate form:

$$\sum_{i \in V_d} x_{ik} \leq (n - p + 1) y_k \quad \forall k \in V_h, \tag{2.8}$$

which specify that if a hub has been located on node $k \in V_h$, then at most $n - p + 1$ demand nodes can be assigned to it.

The quadratic component (2.2) of the objective function can be linearised using standard techniques, one of which is to replace the expression $x_{ik} x_{jm}$ by a new binary variable z_{ijkm} for all $i, j \in V_d$, $i \neq j$ and $k, m \in V_h$ using the additional constraints mentioned here:

$$z_{ijkm} \leq x_{ik} \tag{2.9}$$

$$z_{ijkm} \leq x_{jm} \tag{2.10}$$

$$z_{ijkm} \geq x_{ik} + x_{jm} - 1, \tag{2.11}$$

resulting in an augmented formulation with a linear objective function. In Section 2.3.1.2, we present alternative integer linear programming formulations for the single-allocation p-hub median problem.

2.3.1.2 Integer linear programming formulations

The formulation presented in this section uses the same assignment variables x_{ik} and location variables y_k. Additionally, a continuous variable z_{ijkm} is defined to represent the fraction of the (directed) flow from origin $i \in V_d$ to destination $j \in V_d \backslash \{i\}$ that is routed via hubs $k, m \in V_h$ in the given order, with a cost denoted by $\hat{c}_{ijkm} = c_{ik} + \alpha c_{km} + c_{mj}$.

$$\text{Minimise} \sum_{i,j \in V_d; i \neq jk, m \in V_h; k \neq m} \sum w_{ij} \hat{c}_{ijkm} z_{ijkm} \tag{2.12}$$

subject to

$$\sum_{k \in V_h} y_k = p \tag{2.13}$$

$$\sum_{k,m \in V_h} z_{ijkm} = 1 \quad \forall i, j \in V_d, i \neq j \tag{2.14}$$

$$\sum_{j \in V_d, m \in V_h} (w_{ij} z_{ijkm} + w_{ji} z_{jimk}) = \sum_{j \in V_d} (w_{ij} + w_{ji}) x_{ik} \quad \forall i \in V_d, \tag{2.15}$$

$$\forall k \in V_h$$

$$x_{ik} \leq y_k \quad \forall i \in V_d, k \in V_h \tag{2.16}$$

$$x_{ik} \in \{0, 1\} \quad \forall (i, k) \in E \tag{2.17}$$

$$y_k \in \{0, 1\} \quad \forall i, k \in V_h \tag{2.18}$$

$$z_{ijkm} \geq 0 \quad \forall i, j \in V_d, i \neq j, k, m \in V_h. \tag{2.19}$$

This formulation minimises the total cost of flow through the objective function (2.12) under a requirement (2.13) that exactly p hubs are located. Constraints (2.14) state that the total flow for an OD pair (i, j) should be routed from its origin to destination, but which alone do not guarantee the single assignment restriction. To enforce this, constraints (2.15) are introduced, which state that if an assignment $x_{ik} = 1$ is made for a particular demand node $i \in V_d$ and a hub $k \in V_h$, then the total flow out of and into node i should flow on the link (i, k), implying that no other hub but k can be used for the assignment. The remaining constraints are as in Section 2.3.1.1.

If z_{ijkm} are defined as binary variables indicating whether or not the flow from i to j flows through hubs k and m in the given order, then constraint (2.15) can be written in a simplified form as follows:

$$x_{ik} + x_{jm} - 2z_{ijkm} \geq 0, \tag{2.20}$$

which enforce the condition that if $z_{ijkm} = 1$, then it must be that $x_{ik} = x_{jm} = 1$ in any feasible solution.

Example 2.1

Consider the hub location instance shown in Figure 2.3 defined on a grid composed of unit squares. The hub locations are represented by the square nodes $V_h = \{1, 2, 3, 4\}$. Similarly, $V_d = \{5, \ldots, 8\}$ is the index set of origin or demand nodes. We assume, for the purposes of this illustrative example, that a unit of flow is to be shipped from node 5 to node 8 ($w_{58} = 1$), one unit from node 6 to node 8 ($w_{68} = 1$) and one unit from node 5 to node 7 ($w_{57} = 1$). The distances are given by the length of the path between an individual pair of nodes, and the discount rate is $\alpha = 0.4$. This means, for example, that the unit cost of flow using direct shipment from supply node 5 to demand node 8 is $\sqrt{4^2 + 5^2} = 6.4$, whereas shipping a flow from node 5 to node 8 via the hub pair $(1, 4)$ will imply a unit cost $1 + 5 \times 0.4 + 1 = 4$ units due to the economies of scale. Assume also that there are four existing but inactive hubs, one on each node $V_h = \{1, 2, 3, 4\}$, and the decision is to select $p = 2$ two hubs to activate, which implies that the fixed cost of using a hub is negligible for this example.

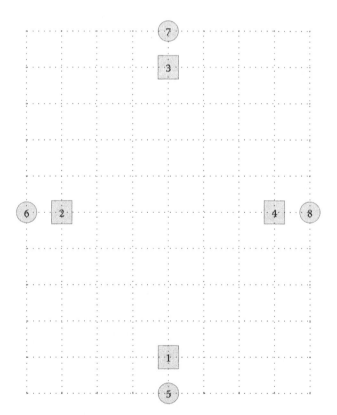

Figure 2.3 The hub location instance described in Example 2.1.

To be able to write the objective function of the integer linear programming formulation of this problem, one would have to calculate the cost of routing the flow for each origin–destination pair using a pair of hubs, similar to this calculation. One would also have to bear in mind that the variables are defined on the basis of a directed network. For example, whereas the variable z_{5814} indicates an assignment of node 5 to hub 1 and node 8 to hub 4 with a unit cost of 4, the variable z_{5841} implies an assignment of node 5 to hub 4 and node 8 to hub 1, implying a unit flow cost of $5.83 + 5 \times 0.4 + 5.65 = 13.48$. The complete objective function is given here:

Minimise

$$+ 6.66z_{5811} + 12.83z_{5822} + 14.66z_{5833} + 6.83z_{5844}$$

$$+ 11.32z_{6811} + 8z_{6822} + 11.32z_{6833} + 8z_{6844}$$

$$+ 10z_{5711} + 11.66z_{5722} + 10z_{5733} + 11.66z_{5744}$$

$$+ 6.4z_{5812} + 9.85z_{5813} + 4z_{5814} + 13.48z_{5823}$$

$$+ 9.23z_{5824} + 7.2z_{5834}$$

$$+ 13.48z_{5821} + 13.05z_{5831} + 13.48z_{5841} + 9.6z_{5832}$$

$$+ z_{5842} + 13.48z_{5843}$$

$$+ 11.05z_{6812} + 14.5z_{6813} + 8.65z_{6814} + 8.65z_{6823}$$

$$+ 4.4z_{6824} + 8.65z_{6834}$$

$$+ 8.65z_{6821} + 14.5z_{6831} + 11.05z_{6841} + 11.05z_{6832}$$

$$+ 9.2z_{6842} + 11.05z_{6843}$$

$$+ 8.83z_{5712} + 5.2z_{5713} + 8.83z_{5714} + 8.83z_{5723}$$

$$+ 14.06z_{5724} + 12.43z_{5734}$$

$$+ 12.03z_{5721} + 8.4z_{5731} + 12.03z_{5741} + 12.03z_{5732}$$

$$+ 14.06z_{5742} + 8.83z_{5743}.$$

The objective function accounts for all possible flows between hub pairs for each origin–destination pair, which, for this instance, is three (5–8, 6–8 and 5–7). Constraints (2.13) take the following form:

$$y_1 + y_2 + y_3 = 2.$$

We now move on to modelling the flow for each OD pair, which, for this instance, will be ensured through the following constraints:

$$z_{5811} + z_{5822} + z_{5833} + z_{5844}$$

$$+ z_{5812} + z_{5813} + z_{5814} + z_{5823} + z_{5824} + z_{5834}$$

$$+ z_{5821} + z_{5831} + z_{5841} + z_{5832} + z_{5842} + z_{5843} = 1$$

$$z_{6811} + z_{6822} + z_{6833} + z_{6844}$$

$$+ z_{6812} + z_{6813} + z_{6814} + z_{6823} + z_{6824} + z_{6834}$$

$$+ z_{6821} + z_{6831} + z_{6841} + z_{6832} + z_{6842} + z_{6843} = 1$$

$$z_{5711} + z_{5722} + z_{5733} + z_{5744}$$

$$+ z_{5712} + z_{5713} + z_{5714} + z_{5723} + z_{5724} + z_{5734}$$

$$+ z_{5721} + z_{5731} + z_{5741} + z_{5732} + z_{5742} + z_{5743} = 1.$$

The single assignment constraints are written for each origin or destination node in set V_d and a potential hub location. Here, we present one of these constraints in their explicit form for $i = 5$ and $k = 1$ of this instance:

$$z_{5711} + z_{5712} + z_{5713} + z_{5714} + z_{5811} + z_{5812} + z_{5813} + z_{5814}$$

$$+ z_{7521} + z_{7531} + z_{7541} + z_{8521} + z_{8531} + z_{8541} = 2x_{51},$$

with the interpretation that if node 5 is assigned to hub 1 (i.e. $x_{51} = 1$), then a total of 2 units must flow from node 5 to node 1, which will then be forwarded to one of the other hubs 2, 3 or 4.

The final set of constraints needed for this instance, written for node 1 as a potential hub location, is as follows:

$$x_{51} \leq y_1$$

$$x_{61} \leq y_1$$

$$x_{71} \leq y_1.$$

Similar constraints can be written for hub nodes 2, 3 and 4.

An alternative and tighter formulation for the problem which assumes that a hub can be allocated to itself by using variables $z_{kk} = y_k$ for all $k \in V$ is presented here:

$$\text{Minimise} \quad \sum_{i,j \in V_d; i \neq jk, m \in V_h} \sum w_{ij} \hat{c}_{ijkm} z_{ijkm} \qquad (2.21)$$

subject to

$$\sum_{k \in V_h} x_{kk} = p \qquad (2.22)$$

$$\sum_{k \in V_h} x_{ik} = 1 \qquad \forall i \in V_d \qquad (2.23)$$

$$x_{ik} \leq x_{kk} \qquad \forall i \in V_d, k \in V_h \tag{2.24}$$

$$\sum_{m \in V_h} z_{ijkm} = x_{ik} \qquad \forall i, j \in V_d, i \neq j, k \in V_h \tag{2.25}$$

$$\sum_{k \in V_h} z_{ijkm} = x_{jm} \qquad \forall i, j \in V_d, i \neq j, m \in V_h \tag{2.26}$$

$$z_{ijkm} \geq 0 \qquad \forall i \in V_d, k \in V_h \tag{2.27}$$

$$x_{ik} \in \{0, 1\} \quad \forall (i, k) \in E. \tag{2.28}$$

In this formulation, constraints (2.22) prescribe the number p of hubs to use and function in the same way as constraints (2.13). Constraints (2.23) model the single-allocation restriction by assigning each node in V_d to exactly one hub. Constraints (2.24) allow an allocation of a node $i \in V_d$ to a hub $k \in V_h$ only if a hub is located on node k, in the same way as constraints (2.16). Finally, constraints (2.25) and (2.26) ensure that there is consistency between the flows from origins to destinations, and hub assignments.

Example 2.2

In the following, we present the full set of constraints for the instance described in Example 2.1. The objective function is the same as shown earlier:

$$x_{11} + x_{22} + x_{33} = 2$$

$$x_{51} + x_{52} + x_{53} + x_{54} = 1$$

$$x_{61} + x_{62} + x_{63} + x_{64} = 1$$

$$x_{71} + x_{72} + x_{73} + x_{74} = 1$$

$$x_{81} + x_{82} + x_{83} + x_{84} = 1$$

$$x_{51} \leq x_{11}$$

$$x_{52} \leq x_{22}$$

$$x_{53} \leq x_{33}$$

$$x_{54} \leq x_{44}$$

$$x_{61} \leq x_{11}$$

$$x_{62} \leq x_{22}$$

$$x_{63} \leq x_{33}$$

$$x_{64} \leq x_{44}$$

$$x_{71} \leq x_{11}$$

$$x_{72} \leq x_{22}$$

$$x_{73} \leq x_{33}$$

$$x_{74} \leq x_{44}$$

$$x_{81} \leq x_{11}$$

$$x_{82} \leq x_{22}$$

$$x_{83} \leq x_{33}$$

$$x_{84} \leq x_{44}$$

$$z_{5811} + z_{5812} + z_{5813} + z_{5814} = x_{51}$$

$$z_{5821} + z_{5822} + z_{5823} + z_{5824} = x_{52}$$

$$z_{5831} + z_{5832} + z_{5833} + z_{5834} = x_{53}$$

$$z_{5841} + z_{5842} + z_{5843} + z_{5844} = x_{54}$$

$$z_{6811} + z_{6812} + z_{6813} + z_{6814} = x_{61}$$

$$z_{6821} + z_{6822} + z_{6823} + z_{6824} = x_{62}$$

$$z_{6831} + z_{6832} + z_{6833} + z_{6834} = x_{63}$$

$$z_{6841} + z_{6842} + z_{6843} + z_{6844} = x_{64}$$

$$z_{5711} + z_{5712} + z_{5713} + z_{5714} = x_{51}$$

$$z_{5721} + z_{5722} + z_{5723} + z_{5724} = x_{52}$$

$$z_{5731} + z_{5732} + z_{5733} + z_{5734} = x_{53}$$

$$z_{5741} + z_{5742} + z_{5743} + z_{5744} = x_{54}$$

$$z_{5811} + z_{5821} + z_{5831} + z_{5841} = x_{81}$$

$$z_{5812} + z_{5822} + z_{5832} + z_{5842} = x_{82}$$

$$z_{5813} + z_{5823} + z_{5833} + z_{5843} = x_{83}$$

$$z_{5814} + z_{5824} + z_{5834} + z_{5844} = x_{84}$$

$$z_{6811} + z_{6821} + z_{6831} + z_{6841} = x_{81}$$

$$z_{6812} + z_{6822} + z_{6832} + z_{6842} = x_{82}$$

$$z_{6813} + z_{6823} + z_{6833} + z_{6843} = x_{83}$$

$$z_{6813} + z_{6823} + z_{6843} + z_{6844} = x_{84}$$

$$z_{5711} + z_{5721} + z_{5731} + z_{5741} = x_{71}$$

$$z_{5712} + z_{5722} + z_{5732} + z_{5742} = x_{72}$$

$$z_{5713} + z_{5723} + z_{5733} + z_{5743} = x_{73}$$

$$z_{5714} + z_{5724} + z_{5734} + z_{5744} = x_{74}.$$

The full set of constraints corresponding to the nonnegativity and integrality requirements on the variables is not shown here explicitly.

2.3.2 Multiple-allocation p-hub median problem

Allowing demand nodes to be served by more than one hub gives rise to the *multiple-allocation p-hub median problem*. One way to model the problem is presented here, using the same variables as in the previous section.

$$\text{Minimise} \quad \sum_{i,j \in V_d; i \neq jk, m \in V_h} \sum w_{ij} \hat{c}_{ijkm} z_{ijkm} \tag{2.29}$$

subject to

$$\sum_{k \in V_h} y_k = p \tag{2.30}$$

$$\sum_{k,m \in V_h} z_{ijkm} = 1 \qquad \forall i,j \in V_d, i \neq j \tag{2.31}$$

$$z_{ijkm} \leq y_k \qquad \forall i,j \in V_d, i \neq j, k, m \in V_h \tag{2.32}$$

$$z_{ijkm} \leq y_m \qquad \forall i,j \in V_d, i \neq j, k, m \in V_h \tag{2.33}$$

$$z_{ijkm} \geq 0 \qquad \forall i \in V_d, k \in V_h \tag{2.34}$$

$$y_k \in \{0, 1\} \qquad \forall i, k \in V_h. \tag{2.35}$$

In this formulation, constraints (2.31) ensure that OD flows are routed using hubs. Constraints (2.32) and (2.33) model the situation that, for a flow to be able to use hubs on nodes k, m, hubs must be located on these nodes. Although this formulation is written on the basis that the z_{ijkm} variables can take fractional values, the absence of any capacity constraints on the links actually implies that all these variables will be equal to either zero or one. The reason is that each flow between a node pair ($i \in V_d, j \in V_d$) will be routed on the least cost path from origin to destination. However, the formulation still allows multiple assignments as the least cost paths for two different OD pairs $(i \in V_d, j_1 \in V_d)$ and $(i \in V_d, j_2 \in V_d \backslash \{j_2\})$ might be different.

It is possible to strengthen the formulation by replacing inequalities (2.32) and (2.33) by the following inequalities:

$$\sum_{m \in V_h} z_{ijkm} \leq y_k \quad \forall i, j \in V_d, i \neq j, k \in V_h \tag{2.36}$$

$$\sum_{k \in V_h} z_{ijkm} \leq y_m \quad \forall i, j \in V_d, i \neq j, m \in V_h. \tag{2.37}$$

It is easy to see that any solution to (2.36) and (2.37) satisfies inequalities (2.32) and (2.33), but the opposite is not necessarily true, giving way to tighter linear programming bounds.

A further improvement can be made on the formulation by combining inequalities (2.36) and (2.37) in a non-trivial fashion as follows:

$$\sum_{m \in V_h} z_{ijkm} + \sum_{m \in V_h} z_{ijmk} \leq y_k \quad \forall i, j \in V_d, i \neq j, k \in V_h. \tag{2.38}$$

Inequalities (2.38) state that if a flow originating from node $i \in V_d$ and ending at node $j \in V_d$ passes through a node $k \in V_h$, then there needs to be a hub located on node k which forces $y_k = 1$.

Example 2.3

Assume that there is an additional demand point in the instance described in Example 2.1, which is shown by node 9 in Figure 2.4. There is the need to send one unit of flow from node 9 to node 7, and a further unit of flow from node 9 to node 8. If a single-assignment restriction is imposed, node 9 can be assigned to hub 1 and the flows can be routed such that $z_{9713} = 1$ and $z_{9814} = 1$, with the former assignment incurring a cost equal to $\hat{c}_{9713} = c_{91} + 0.4c_{13} + c_{37} = 3 + 3.2 + 1 = 7.2$ and the latter incurring a cost equal to 6, implying a total cost of 13.2. However, if multiple assignments were allowed, then it would be possible to route one unit of flow from node 9 to node 7 via hubs 2 and 3 $(z_{9723} = 1)$, and another unit of flow from 9 to node 8 via hubs 1 and 4 $(z_{9814} = 1)$, implying a total cost of 13 units. The latter solution is shown in Figure 2.4 using the solid links. This small example shows that allowing multiple allocations may reduce the overall cost, but one must bear in mind that this is also dependent on the values of parameters such as α and whether there are any fixed costs to take into account for locating or using additional hubs.

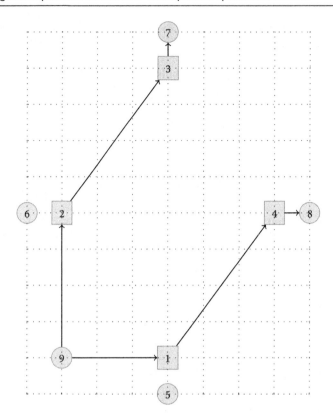

Figure 2.4 The hub location instance described in Example 2.3.

2.4 Capacitated hub location problems

In situations where hubs can only process a limited amount of flow, one is concerned with determining the number and location of hubs where the flow into a hub is constrained by its capacity. The objective is to minimise the total cost of flows and hubs, giving rise to the *capacitated hub location problem*. In this case, the requirement that exactly p hubs should be located is relaxed as certain values of p might render the problem infeasible. Instead, one considers a fixed installation cost g_k for locating a hub on node $k \in V_h$ and an additional constraint that the total flow into this hub should not exceed the fixed capacity π_k of the hub.

A mixed integer programming formulation for this version of the problem is presented in the following:

$$\text{Minimise} \sum_{i,j \in V_d; i \neq jk, m \in V_h} \sum w_{ij} \hat{c}_{ijkm} z_{ijkm} + \sum_{k \in V_h} g_k x_{kk} \qquad (2.39)$$

subject to

$$\sum_{k,m\in V_h} z_{ijkm} = 1 \qquad \forall i,j \in V_d, i \neq j \tag{2.40}$$

$$x_{ik} \leq x_{kk} \qquad \forall i \in V_d, k \in V_h \tag{2.41}$$

$$\sum_{m\in V_h} z_{ijkm} = x_{ik} \qquad \forall i,j \in V_d, i \neq j, k \in V_h \tag{2.42}$$

$$\sum_{k\in V_h} z_{ijkm} = x_{jm} \qquad \forall i,j \in V_d, i \neq j, m \in V_h \tag{2.43}$$

$$\sum_{i,j\in V_d, i\neq j} w_{ij}x_{ik} \leq \pi_k x_{kk} \qquad \forall k \in V_h \tag{2.44}$$

$$z_{ijkm} \geq 0 \qquad \forall i \in V_d, k \in V_h \tag{2.45}$$

$$x_{ik} \in \{0,1\} \qquad \forall(i,k) \in E. \tag{2.46}$$

In this formulation, constraints (2.44) model the additional requirement that the total flow sent to a node k should not exceed the capacity if a hub is located on this node. All other constraints are as explained in the previous sections.

One other interesting variant of the capacitated hub location problem is where the hub capacities are not fixed but are themselves decision variables. One way to consider such decisions is to assume a discrete set $U = \{1,2,\ldots\}$ of capacity levels, where each level $l \in U$ corresponds to a certain capacity π_l and a fixed installation cost g_{kl} for a possible hub node k. In this case, a new decision variable y_{kl} is used, which is equal to 1 if a hub on a node $k \in V_h$ is located with a capacity at level $l \in U$, and 0 otherwise. This problem can be formulated in the same way as (2.39) through (2.46), but the objective function (2.39) of the earlier formulation should be changed as follows:

$$\text{Minimise} \sum_{i,j\in V_d; i\neq j k,m\in V_h} w_{ij}\hat{c}_{ijkm}z_{ijkm} + \sum_{k\in V_h, l\in U}\sum g_{kl}y_{kl}. \tag{2.47}$$

Furthermore, constraints (2.44) should be changed to the following:

$$\sum_{i,j\in V_d, i\neq j} w_{ij}x_{ik} \leq \sum_{l\in U}\pi_{kl}y_{kl} \qquad \forall k \in V_h, \tag{2.48}$$

where the right-hand-side limits the flow into a hub located at node k to the capacity corresponding to level $l \in U$ selected. Finally, the following constraints are needed to model the relationship between the location and capacity selection variables,

$$\sum_{l \in U} y_{kl} = x_{kk} \quad \forall k \in V_h, \tag{2.49}$$

which state that if a hub is located at a node $k \in V_h$, indicating that $x_{kk} = 1$, then the hub can only be assigned one of the capacity levels in the set U, implying that there should only be one $l^* \in U$ for which $y_{kl^*} = 1$ in any feasible solution of the problem.

2.5 Other types of hub location problems

This section describes two other types of hub location problems that mainly arise in the distribution of time-sensitive freight, namely the p-hub centre problem and the hub covering problem. These two problems differ from the traditional hub location problems hitherto discussed in the way that they account for cost. In particular, the concept of coverage is used in these types of problems. Further details are given here.

2.5.1 p-Hub centre problem

Hub centre problems arise in situations where one wishes to avoid sending flows over paths that are too costly, where the latter can be measured with respect to time or the actual operational costs. One context in which such restrictions arise is the distribution of perishable goods, such as seafood, floral products and fruits, which should be shipped to the destination (e.g. consumers, retailers) as quickly as possible to maintain the product freshness and quality at the point of arrival. In this case, the objective of simply minimising the total costs may no longer be suitable as the resulting solutions may still contain paths that are too long in duration or too high in cost. Instead, a more suitable objective is to minimise the maximum time spent in transporting commodities between any origin and destination pair, giving rise to what is known as the p-hub centre problem. In the remainder of this section, we assume that the problem allows multiple allocations, but the formulations can easily be adapted for the case of a single allocation.

One way in which the p-hub centre problem can be formulated is similar to that of the p-hub median problem, in particular using constraints (2.13) through (2.19), but with a different objective function reflecting the nature of the problem, as given here:

$$\text{Minimise} \quad \max_{i,j \in V_d; i \neq j, k, m \in V_h} \hat{c}_{ijkm}\{z_{ijkm}\}, \tag{2.50}$$

which is a nonlinear expression. A way of linearising the expression (2.50) is by using an auxiliary variable ξ as follows:

Minimise ξ (2.51)

subject to

$$\xi \geq z_{ijkm}\hat{c}_{ijkm} \quad \forall i,j \in V_d, i \neq j, k,m \in V_h \tag{2.52}$$

$$z_{ijkm} \in \{0,1\} \quad \forall i,j \in V_d, i \neq j, k,m \in V_h, \tag{2.53}$$

where ξ represents the largest of the values $\hat{c}_{ijkm}z_{ijkm}$ over all $i,j \in V_d; i \neq j$ and $k,m \in V_h$. The formulation also contains the following constraints of the p-hub median problem.

$$\sum_{k \in V_h} y_k = p$$

$$\sum_{k,m \in V_h} z_{ijkm} = 1 \qquad\qquad \forall i,j \in V_d, i \neq j$$

$$\sum_{j \in V_d, m \in V_h} \left(w_{ij}z_{ijkm} + w_{ji}z_{jimk}\right) = \sum_{j \in V_d}(w_{ij} + w_{ji})x_{ik} \quad \forall i \in V_d, k \in V_h$$

$$x_{ik} \leq y_k \qquad\qquad \forall i \in V_d, k \in V_h$$

$$x_{ik} \in \{0,1\} \qquad\qquad \forall (i,k) \in E$$

$$y_k \in \{0,1\} \qquad\qquad \forall i,k \in V_h.$$

It is important to note here that the z_{ijkm} variables are now required to be binary with the new objective function.

The following example illustrates the effect that the objective function (2.50) may have on the solution of a hub location problem, in particular when solving a p-hub median and p-hub centre problem on a given instance.

Example 2.4

Consider a hub location instance, as shown in Figure 2.5, with three candidate hub location sites represented by the node set $V_h = \{1,2,3\}$ and four origin/destination points as shown in the node set $V_d = \{4,5,6,7\}$. The network on which the instance is defined is not complete. The only links that can be used are those shown in the figure, where the unit shipment times (in hours) are as indicated on the links. The discount factor to use on the inter-hub links is $\alpha = 0.6$.

A fresh produce supplier wishes to use $p = 2$ hubs on the node set $V_h = \{1,2,3\}$ to regularly ship their products to its customers located on nodes $\{6,7\}$ from their warehouses located on nodes $\{4,5\}$. In particular,

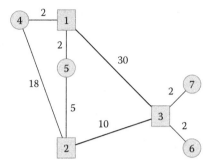

Figure 2.5 The hub location instance described in Example 2.4.

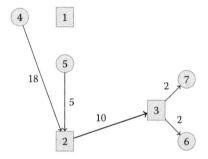

Figure 2.6 The solution to the p-hub median problem for Example 2.4.

one container needs to be shipped from node 4 to node 6, and another container is to be sent from node 5 to node 7 everyday. An investment has already been made for two hub facilities to be located, for which reason the location cost will not need to be taken into account. The producer would like to design the network in such a way so as to be able to offer a guaranteed service time of at most 24 hours for any of the two shipments.

If the location selection problem depends on minimising the total shipment time, then a p-hub median problem will need to be solved for this instance. The resulting solution is as shown in Figure 2.6, where the two hubs are located on nodes 2 and 3, where nodes 4 and 5 are assigned to the hub at node 2. Similarly, nodes 6 and 7 are supplied from the hub at node 3. In this solution, one container follows the path 4–2–3–6 with a shipment time equal to $\hat{c}_{4623} = 18 + 10 \times 0.6 + 2 = 26$ hours. The other container follows the path 5–2–3–7 with a shipment time equal to $\hat{c}_{5723} = 5 + 10 \times 0.6 + 2 = 13$ hours. The total shipment time for the instance is $26 + 13 = 39$ hours, where $\max\{\hat{c}_{4623}, \hat{c}_{5723}\} = 26$ hours.

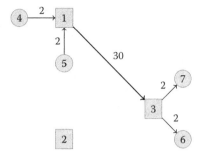

Figure 2.7 The solution to the *p*-hub centre problem for Example 2.4.

However, the shipper is not entirely satisfied with the solution, given that one container takes 26 hours to reach its destination, whereas the other requires 13 hours, and would like to achieve a better balance between the time taken for the two shipments. For this reason, it is more appropriate to solve a *p*-hub centre problem on this instance, the solution of which is shown in Figure 2.7. As this solution shows, the two hubs are now located on nodes 1 and 3, where the assignment of nodes 4 and 5 has changed to the hub at node 1. Nodes 6 and 7 are still supplied from the hub at node 3. In this solution, the container shipped from node 4 now follows the path 4–1–3–6 with a shipment time equal to $\hat{c}_{4613} = 2 + 10 \times 0.6 + 2 = 22$ hours. The container shipped from node 5 follows the path 5–1–3–7 with the same shipment time, namely 22 hours. In this case, while the total shipment time has increased to $22 + 22 = 44$ hours, the maximum shipment time has been reduced down to $\max\{\hat{c}_{4613}, \hat{c}_{5713}\} = 22$ hours. The new solution also meets the desired service level as both shipments would, in this case, be arriving at their destination within 24 hours of leaving their origin.

The use of the ξ variable is a standard way of linearising objective functions of the type min max. Other formulations can be described for the *p*-hub centre problem by adapting the quadratic integer programming formulation described earlier for the *p*-hub median problem (2.1) through (2.7) in the following way:

Minimise ξ,   (2.54)

subject to

$$\sum_{k \in V_h} y_k = p \qquad\qquad (2.55)$$

$$\sum_{k \in V_h} x_{ik} = 1 \qquad\qquad \forall i \in V_d \qquad\qquad (2.56)$$

$$x_{ik} \leq y_k \qquad\qquad \forall i \in V_d, k \in V_h \qquad\qquad (2.57)$$

$$\xi \geq (c_{ik} + \alpha c_{km} + c_{jm})x_{ik}x_{jm} \quad \forall i,j \in V_d, i \neq j, k, m \in V_h \qquad (2.58)$$

$$x_{ik} \in \{0,1\} \qquad\qquad \forall (i,k) \in E \qquad\qquad (2.59)$$

$$y_k \in \{0,1\} \qquad\qquad \forall i, k \in V_h, \qquad\qquad (2.60)$$

where the quadratic constraint (2.58) along with the objective function (2.54) is a linearisation of the following objective function:

$$\text{Minimise} \max_{i,j \in V_d; i \neq j, k, m \in V_h} (c_{ik} + c_{km} + c_{jm})x_{ik}x_{jm}. \qquad (2.61)$$

There are two known non-trivial ways of linearising constraints (2.58):

1. The following linear constraints correctly linearise constraints (2.58):

$$\xi \geq (c_{ik} + \alpha c_{km})x_{ik} + c_{jm}x_{jm} \quad \forall i,j \in V_d, i \neq j, k, m \in V_h \qquad (2.62)$$

This follows from the fact that if $x_{ik} = x_{jm} = 1$, then constraints (2.58) and (2.62) both impose the same bound on the ξ variable. The same applies for the case $x_{ik} = x_{jm} = 0$. If, however, $x_{ik} = 1$ and $x_{jm} = 0$, then constraints (2.58) read $\xi \geq 0$, whereas constraints (2.62) impose a lower bound $c_{ik} + \alpha c_{km}$ on ξ. However, this bound will be dominated by $c_{ik} + \alpha c_{km} + c_{tm}$ for $t \in V_d \backslash \{j\}$ due to the assignment constraints (2.56). A similar result can be shown for the case $x_{ik} = 0$ and $x_{jm} = 1$.

2. An alternative linearisation is achieved by the use of additional non-negative and continuous variables r_k, defined as the radius of a hub located on node $k \in V_h$. In particular, the radius is the maximum cost between hub k and the nodes assigned to it, that is $r_k = \max_{i \in V_d : x_{ik}=1} c_{ik}$. Using this definition, it is possible to write the following set of constraints as a linearisation of (2.58):

$$r_k \geq c_{ik}x_{ik} \qquad\qquad \forall i \in V_d, k \in V_h \qquad\qquad (2.63)$$

$$\xi \geq r_k + r_m + \alpha c_{km} \qquad \forall k, m \in V_h. \qquad\qquad (2.64)$$

It is easy to check this validity of this linearisation.

One other way to define the p-hub centre problem is to look at individual links as opposed to paths connecting origin and destination pairs. There are, in general, two types of links in hub networks: (1) between pairs of hubs and (2) between non-hub nodes (origins and destinations) and hub nodes.

In applications where regular services run between two hub nodes, shorter times are desirable on the inter-hub links. For a certain type of goods that need heating or cooling at hub nodes, the time spent on the links between non-hub and hub nodes cannot be too long. In such cases, the shipment times on individual links become more important and it becomes desirable to minimise the maximum time spent on any link. This problem can be formulated with the following objective:

$$\text{Minimise} \max_{i,j \in V_d; i \neq j, k, m \in V_b} \{\max\{c_{ik}, c_{mj}, \alpha c_{km}\}\}, \tag{2.65}$$

expressed as a nonlinear function but can be linearised using standard techniques similar to the one described earlier.

2.5.2 Hub covering problem

Although the p-hub centre problem discussed in the previous section can be used to make decisions concerning the location of hubs in an attempt to avoid costs (time or distance) that are too high, they do not necessarily guarantee that the costs will be within a predetermined limit. These limits can either reflect quality of service considerations, such as guaranteed shipment times in express deliveries (as was the case with the 24 hours' service time guarantee in Example 2.4), or be dictated by the nature of the goods being shipped, such as perishable or time-sensitive goods. In other cases such as postal services, a hub might only serve customers located within a certain radius of its location. The objective in such cases is to locate hubs in the network at minimum cost such that demands are covered, giving rise to the *hub covering problem*.

Coverage can be defined for individual nodes or for origin–destination pairs, for which it is useful to introduce the concept of *cover radius*, which can be defined individually for a node $i \in V_d$ and represented as β_i, or jointly for an origin–destination pair (i, j) and represented as β_{ij}. One can then define coverage for an origin–destination pair (i, j) and a hub pair (k, m) as follows:

C1. If $c_{ik} \leq \beta_i$ and $c_{jm} \leq \beta_j$,
C2. If $c_{ik} + c_{mj} + \alpha c_{km} \leq \beta_{ij}$,
C3. If $\max\{c_{ik}, c_{mj}, \alpha c_{km}\} \leq \beta_{ij}$,

where the first definition assumes that nodes i and j are covered by hubs k and m, if the individual assignment costs are within the respective coverage radii, as per the conditions stated in C1. The second and third definitions follow from the min-sum and min-max objectives found in the p-hub median and p-hub centre problems, respectively.

In this problem, the cost to be minimised is the sum of the individual location costs F_k for each node $k \in V_h$ where a hub is to be located. If $F_k = 1$ for all $k \in V_h$, the problem simply minimises the number of hubs located in the network. The constraints are as in the p-hub median problem, where the problem might be defined with respect to either single or multiple allocation. In the following, we focus on the version with the single-allocation restriction.

Irrespective of which criterion is being used to define coverage, one can use a binary parameter a_{ijkm} that is equal to 1 if an origin–destination pair (i, j) is covered by a hub pair (k, m), and 0 otherwise. The values of a_{ijkm} can be computed by checking conditions C1–C3. Using this definition, a general integer programming formulation for the hub covering problem can be presented as follows:

$$\text{Minimise} \sum_{k \in V_h} g_k y_k \tag{2.66}$$

subject to

$$\sum_{k \in V_h} y_k = p \tag{2.67}$$

$$\sum_{k \in V_h} x_{ik} = 1 \qquad \forall i \in V_d \tag{2.68}$$

$$\sum_{k,m \in V_h} a_{ijkm} x_{ik} x_{jm} \geq 1 \qquad \forall i, j \in V_d, i \neq j \tag{2.69}$$

$$x_{ik} \leq y_k \qquad \forall i \in V_d, k \in V_h \tag{2.70}$$

$$x_{ik} \in \{0, 1\} \qquad \forall (i, k) \in E \tag{2.71}$$

$$y_k \in \{0, 1\} \qquad \forall k \in V_h. \tag{2.72}$$

The formulation follows that of the p-median problem where coverage is modelled through the use of quadratic constraints (2.69). A linearisation of these constraints is possible by replacing the quadratic term $x_{ik} x_{jm}$ by a new binary variable z_{ijkm}, as in the case of the p-hub median problem.

For the coverage defined by C2, an alternative quadratic expression that can be used in lieu of constraints (2.69) is as follows:

$$(c_{ik} + \alpha c_{km} + c_{jm}) x_{ik} x_{jm} \leq \beta_{ij}. \tag{2.73}$$

In this case, a simpler linearisation is obtained by replacing constraints (2.73) by the following inequalities:

$$(c_{ik} + \alpha c_{km}) x_{ik} + c_{jm} x_{jm} \leq \beta_{ij}, \tag{2.74}$$

and it is easy to check that these inequalities correctly linearise constraints (2.73) as in the p-hub centre problem.

It is possible to strengthen the formulation by using the following additional inequalities:

$$x_{ik} + x_{jm} \leq 1, \tag{2.75}$$

written for all $i, j \in V_d$, $i \neq j$, and $k, m \in V_h$ such that $c_{ik} + \alpha c_{km} + c_{jm} > \beta_{ij}$. In fact, a set of such inequalities of the type (2.75) written for (i, j, k) and $m^1, \dots, m^T \in V_h$, one can aggregate them into the following stronger inequalities:

$$x_{ik} + \sum_{t=1}^{T} x_{jm^t} \leq 1, \tag{2.76}$$

which is valid since the assignment constraints (2.68) imply $\displaystyle\sum_{t=1}^{T} x_{jm^t} \leq 1$.

2.6 Congestion in hub location problems

While the use of hubs enables shippers or carriers to take advantage of economies of scale, which in turn encourages the use of hubs as points of consolidation within transportation networks, increased amounts of flow sent in or processed by a hub might be a cause for congestion at such nodes. Congestion is one of the causes for delay at freight hubs, for example rail yards, ports or even cities. In rail transport, for example, it is well known that freight cars, in a rail network, spend most of their time in terminals or classification yards. This is due to the fact that the same facility has to be used for consolidation and classification operations for a variety of vehicles carrying different types of freight. In these yards, cars usually go through the following operations: inspection, classification, assembly, accumulation and connection. The classification process constitutes the fundamental source of delay in the terminals, and this increases with the amount of classification, which is correlated with the number of cars to be classified. In airports, an increased number of plane arrivals over a given period of time in a day may result in queues, again a natural consequence of the congestion phenomenon. Similarly, the number of ship arrivals into a port on a given day might give rise to congestion, either in the berthing area or within the storage area of the port (e.g. where containers are stacked).

Imposing a simple capacity restriction on the amount of flow that each hub would accommodate over a given period of time, as was done in Section 2.4, would be one way to ensure that the hub does not receive more flows that it is able to process. However, this approach does not necessarily capture the effect of congestion. Congestion is known to increase exponentially with the amount of flow and is better captured through the use of convex functions. In particular, let f_k denote the amount of flow sent to a hub $k \in V_h$, which, in terms of the decision variables of the hub location problem introduced earlier, can be stated as follows:

$$f_k = \sum_{i,j \in V_d; i \neq jm \in V_h; k \neq m} w_{ij} z_{ijkm}. \tag{2.77}$$

One way to capture the effect of congestion is to use the following function for each hub $k \in V_h$:

$$a(f_k)^b, \tag{2.78}$$

defined for amounts of flow exceeding a certain threshold ϵ. Here, a and $b \geq 1$ (but normally $b > 1$) are parameters that define the shape of the function and are used to capture the severity of congestion. For example, the higher the value of b, the quicker the exponential growth is in congestion. The delay itself can then be defined as a function of congestion, whether it be defined in terms of time, cost or any other unit of measurement. Figure 2.8 shows how the congestion function (2.78) behaves for $a = 1$, and with different values for the parameter b.

Incorporating the congestion function (2.78) into hub location models results in an integer nonlinear programming formulation, which is difficult to solve optimally, although algorithms such as Lagrangean relaxation or generalised Benders decomposition (see Chapter 8) have been successfully

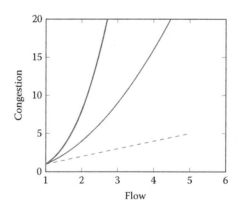

Figure 2.8 Capturing the effect of congestion with $a = 1$ and $b = 1$ (dashed line) and $b = 2$ (line), $b = 3$ (thick line).

used to solve such formulations. The papers cited in the 'References and further readings' section at the end of this chapter provide some pointers to the relevant literature. Existing studies also provide numerical evidence that incorporating congestion into hub location results in a better balanced distribution of flows, and the location decisions are affected by the changes in the values of the parameters a and b. Furthermore, it has been suggested that significant congestion within a hub-and-spoke network reduces its effectiveness and the ability of providing cost savings.

2.7 Facility location problems

A special case of hub location problems arises when there are assumed to be no interactions, and therefore no links, between hub nodes, which prohibit any flows between pairs of hubs. In this case, all flows are assumed to be direct shipments between demand points and one or more hubs. This particular case is known as the *facility location problem*, which assumes a different context where hub nodes are *facilities* (e.g. factories, warehouses, storage points), each of which are origin points, supplying a particular commodity. Facility location problems only involve demand points, which can also be referred to as *customers*, who require a certain amount of this commodity that should be met in full. The facility location problem involves only a subset of the decisions of the hub location problem, namely assignment and location. The flow decisions are often not explicitly stated but are a consequence of the assignment decisions. Further details will be provided here.

A sample facility location problem instance with three facilities and six demand points defined on a grid is given in Figure 2.9. In this instance, the

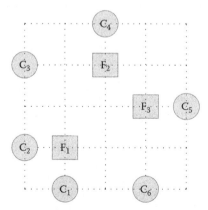

Figure 2.9 A sample instance of the facility location problem.

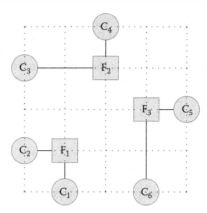

Figure 2.10 A typical solution to a facility location problem.

nodes labelled F_1, F_2 and F_3 are three facilities that are available to supply a commodity to six customers C_1–C_6.

A feasible solution to the instance is given in Figure 2.10, where the assignments are shown through the solid links. In the solution shown, facility F_1 serves customers C_1 and C_2, facility F_2 serves customers C_3 and C_4 and facility F_3 serves customers C_5 and C_6. If the distances shown in the grid are used as a measure of cost, and there are no fixed costs associated with using the facilities, then the solution shown is also optimal as the assignments are made to the closest facilities available.

Hub and facility location problems are interlinked closely. For example, it is easy to see that the solution shown in Figure 2.10 is similar to that in Figure 2.1, with the exception that inter-hub connections do not exist in the former. Indeed, facility location problems do not assume a hub-and-spoke network structure. In this particular example, a single-assignment restriction is assumed, but as we will see in the following, this can be relaxed in the same way as was done for the multiple-allocation hub location problem presented in Section 2.3.2.

Applications of facility location problems arise in many contexts, particularly in distribution logistics. Facilities are points where commodities may be produced, processed (from raw material into partial or fully manufactured goods) or stored. Often, there will be a discrete and a finite set of candidate sites on which facilities may be placed, but this incurs a fixed installation cost. Customers are nodes in the network to which demand is sent using direct shipments. The assignment cost of a customer to an installed facility can be fixed, or modelled as a function of the amount of commodity shipped. We now differentiate between several different variants of the facility location problem:

- *Uncapacitated facility location problem*: There is no limitation on the amount of supply that can be sent from a facility to the customers that have been assigned to this facility.
- *Capacitated facility location problem*: Each facility can only supply a limited amount of commodity to all the customers that have been assigned to it. In this case, the total supply of facilities that can be located must be at least the total demand of customers, so that the problem admits a feasible solution. In some instances, the capacity is defined by the *number* of customers that can be served by each facility.
- *Single-assignment*: Each customer can only be assigned to a single facility. It is obvious that optimal solutions of uncapacitated facility location problems will always have the single-assignment property, as each customer will want to receive commodities from a facility providing the cheapest shipment cost.
- *Multiple-assignment facility location problems*: These problems allow each customer to be assigned to and receive direct shipments from multiple facilities. They are particularly relevant in capacitated problems where the capacity constraints may make it impossible for a facility to meet a given customer's demand in its entirety. In this case, the additional flexibility afforded by multiple assignments allows the customers to be served from multiple facilities, in order that their demands are fully met.

Section 2.7.1 presents formulations for some well-known facility location problems.

2.7.1 Uncapacitated facility location problem

Similar to the notation used for hub location problems, we use V_h to denote the candidate set of facility nodes, as a proper subset of the set V of nodes. The node set V_d corresponds to the set of customers, where each customer on node $i \in V_d$ has demand $q_i > 0$ for a given type of commodity. The set of nodes is such that $V_h \cap V_d = \emptyset$, that is, a customer node cannot be a facility node or vice versa. To formulate the facility location problem, we use a binary variable y_j that equals 1 if a facility is to be installed on a candidate node $j \in V_d$, which incurs a fixed cost shown by F_j. We also use a continuous variable x_{ij} that shows the fraction of the demand each customer $i \in V_d$ receives from a facility $i \in V_h$. The unit cost of shipping demand is shown by $c_{ij} > 0$. A formulation for the uncapacitated facility location problem is presented as follows:

$$\text{Minimise} \sum_{k \in V_h} g_j y_j + \sum_{i \in V_d} \sum_{j \in V_h} q_i c_{ij} x_{ij} \tag{2.79}$$

subject to

$$\sum_{j \in V_h} x_{ij} = 1 \qquad \forall i \in V_d \tag{2.80}$$

$$x_{ij} \leq y_k \qquad \forall i \in V_d, j \in V_h \tag{2.81}$$

$$x_{ij} \geq 0 \qquad \forall i \in V_d, j \in V_h \tag{2.82}$$

$$y_j \in \{0,1\} \qquad \forall j \in V_h. \tag{2.83}$$

In this formulation, the two components of the objective function (2.79) represent the total cost of facility installation and assignment, respectively. These two units of cost are often incompatible with respect to the time interval in which they are measured. In particular, the former is usually defined as the total cost of investment (e.g. construction) made into the facility, whereas the latter is normally for shipments done on a regular basis (e.g. daily or monthly). Facility location relates to strategic decisions, whereas transportation activities are either tactical or operational. In order that the two cost components be made comparable, facility installation costs can be brought down to the same level (e.g. daily) as the transportation costs through a simple multiple year appraisal. To this end, one should consider the length of the planning horizon in which the facility will be assumed to be in operation.

The rest of the formulation includes constraints (2.80) that guarantee assignments such that the demand of each customer will be met in full. Constraints (2.82) prohibit assignments of customers to facilities that have not yet been installed. Constraints (2.82) are the nonnegativity restrictions on the assignment variables. Similarly, constraints (2.83) are the integrality restrictions on the location variables.

The formulation presented in the previous section allows, through the use of the continuous variables x_{ij}, multiple assignments such that a customer can be assigned to a number of facilities. It is easy to enforce the single-assignment restriction on the facility location problem by changing the nature of the assignment variables $x_{ij} \geq 0$ to $x_{ij} \in \{0,1\}$, in which case the model (2.79) through (2.83) becomes a (pure) binary integer programming formulation.

2.7.2 Capacitated facility location problem

In this section, we will cover two types of capacity restrictions, one on the amount of demand that can be shipped from each facility in the multiple-assignment case, and the other on the number of customers assigned to each facility assuming single-assignment.

In the first case, we assume that each facility $j \in V_h$ is able to serve up to Π_j units over a given period of time. The period can be defined in terms of days, months, or years, depending on the context. In this case, it would suffice to include the following set of constraints in formulation (2.79) through (2.83),

$$\sum_{j \in V_d} q_i x_{ij} \leq \Pi_j y_j \quad \forall j \in V_h, \tag{2.84}$$

which ensures that a facility should be located on node $j \in V_h$ (i.e. $y_j = 1$), then the assignment of customers to that facility should be such that the associated demand can be met without exceeding the capacity of the facility. For this version of the problem to admit a feasible solution, it must be guaranteed that $\sum_{j \in V_h} \Pi_j y_j \geq \sum_{i \in V_d} q_i$, namely the total capacity available across the potential facilities must be sufficiently large so as to be able to meet the total demand.

In the case of the single-assignment restriction, where $x_{ij} \in \{0, 1\}$, it is easy to extend the formulation to limit the number Π_j of customers that a facility $j \in V_h$ can accommodate by the inclusion of the following set of constraints:

$$\sum_{i \in V_d} x_{ij} \leq \Pi_j y_j \quad \forall j \in V_h.$$

For the problem to admit a feasible solution under this restriction, one must ensure that $\sum_{j \in V_h} \Pi_j y_j \geq |V_d|$.

2.8 Uncertainty in location problems

All problems and models introduced in this chapter so far have assumed a *deterministic* and *static* set of parameters. In other words, all parameters are assumed to be known with certainty at the time of planning and that they will remain constant during the period in which location plans are to be made and implemented. As with any other decision problem, the parameters of location problems, whether hub or facility, are likely to be subject to uncertainty. The uncertainty may be relevant to the facility and shipment costs, as well as to customer-related parameters, such as their location, whether or not they will actually be requiring any service and, more prevalently, to the amounts of demands themselves.

Demand is subject to more variability than the others, as it depends on a number of factors, not all controllable by the suppliers or shippers. The deterministic assumption behind demand might be relevant for cases where there are long-term contracts made between suppliers (facilities) and customers for regular shipments of a given commodity, and where the amount of shipment is agreed between both parties. However, other contexts where demands are not known with certainty will require a different treatment. This is relevant where customer demand might be affected by change in market conditions or unexpected events (such as weather-related changes in demand for a particular product). In this case, the decision maker will often be provided with a number of possible scenarios, each with a likelihood of occurrence (i.e. a probability), and in which demand is treated as a random variable. The actual values of the demands, what is also referred to as realisation of the random variables, will not be known until after a number of decisions, primarily of a strategic or a tactical nature, have been made.

In such problems, decisions are often made in two stages. The first-stage decisions are those that are made *prior* to the realisation of the random variables. The second-stage decisions are those that can be made *after* the decision maker has full knowledge of the values of the random variables. Such a treatment of problems involving uncertainty is known as *two-stage stochastic programming*, and facility location problems serve as a good example for the application of this method.

For example, the uncapacitated facility location problem as described and formulated in Section 2.7.1 has a natural partitioning of its variables, where decisions concerning location decisions are long term, whereas those pertaining to assignment are shorter term. This implies that, when customer demand is stochastic, it is impractical to wait for the realisation of the random variables to make the facility location decisions. Furthermore, once these decisions are made, they cannot be changed easily in the light of the variability in the demands. This restriction does not necessarily apply, however, to the assignment variables, which can be changed fairly frequently (e.g. on a day-to-day basis), as and when the need arises in line with the demand patterns. For this reason, and in the context of two-stage stochastic programming, it will be appropriate to make the location decisions in the first stage and leave the assignments to be made in the second stage of decision making.

The following example motivates the problem and illustrates the application.

Example 2.5

Consider the instance described in Figure 2.11, which shows two demand points C_1 and C_2, and three potential facility locations F_1, F_2 and F_3 defined on a rectangular grid, where the distances between the various points are indicated on the links. It has been estimated using historical data that each customer will either place an order for 10 containers or will require no deliveries at all. Using historical data, three possible scenarios have been identified, for which the probabilities

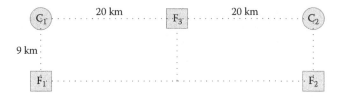

Figure 2.11 The facility location problem instance in Example 2.5.

Table 2.1 Scenarios and probabilities associated with Example 2.5

Demand	Scenario 1	Scenario 2	Scenario 3
C_1	10	0	10
C_2	0	10	10
Probability	0.2	0.2	0.6

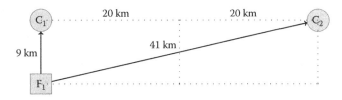

Figure 2.12 Optimal assignments for Example 2.5 when a facility is placed at node F_1.

and the associated demand patterns are summarised in Table 2.1. It is assumed that the location cost of each facility is 200 (monetary) units and that the shipment cost is one unit per kilometre (km) per container. For example, shipping 10 containers of commodity over a distance of 9 km will result in a total cost of 90 (monetary) units.

If, for example, a decision is made to locate a single facility at node F_1, the assignments will be as shown in Figure 2.12, where the arc (F_1, C_1) will be used in Scenarios 1 and 3 and the arc (F_1, C_2) will be used in Scenarios 2 and 3. The expected cost of this solution by using the respective probabilities for each of the three scenarios is calculated as $200 + 0.2 \times 9 \times 10 + 0.2 \times 41 \times 10 + 0.6 \times (41 + 9) \times 10 = 600$ units. Placing a single facility at location at F_2 will incur the same cost.

Now, let us look at locating a single facility at node F_3. At first, this may seem to be counterintuitive, as it would mean having to assign both C_1 and C_2 to facility at node F_3, a facility that is much further away for either customer as compared to F_1 and F_2, respectively. For Scenario 3, for example, the assignments would be as in the solution shown in Figure 2.13, with a cost equal to $200 + 2(20 \times 10) = 600$ units.

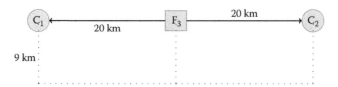

Figure 2.13 Optimal assignments for Example 2.5 when a facility is placed at node F₃.

Figure 2.14 Optimal assignments for Example 2.5 when facilities are placed at nodes F₁ and F₂.

Alternatively, locating two facilities on F_1 and F_2 results in a solution shown in Figure 2.14 with a cost of $200 + 200 + 2(9 \times 10) = 580$ units for the same scenario. However, these calculations favouring the latter solution ignore the stochasticity of demands and do not take into account Scenarios 1 and 2. In fact, the solutions shown in Figures 2.13 and 2.14 have an expected cost equal to 520 and 544 units, respectively, when all three scenarios are taken into account. In the light of this, the former solution is preferable from an uncertainty point of view.

It is impractical to enumerate through all possible solutions for an uncertain facility location problem and to calculate the expected costs, particularly as there will be exponentially many feasible solutions for instances of the problem that are of realistic size. Instead, the uncertainty can be represented by a discrete number of scenarios, each with a certain probability of occurrence, and modelled using integer linear programming as shown in the following.

2.8.1 Scenario-based modelling

Let Ξ be the set of scenarios, where each scenario $\xi \in \Xi$ prescribes a set of demands $q_{i\xi}$ and has a certain probability $p_\xi > 0$ of occurrence. Using the notation defined earlier, we preserve the binary variable $y_j, j \in V_h$ corresponding to location variables for the facilities. We also use a similar assignment variable with the same meaning as before, namely the fraction of demand supplied by facility $j \in V_h$ to customer $i \in V_d$, but this time replicate them for as many scenarios as there are by using the notation $x_{ij\xi}$, which reflects the assignment decisions made under each scenario $\xi \in \Xi$. The objective is to minimise the total cost of locating the facilities, as well as the expected cost of the assignments across all the scenarios. The following is a *deterministic equivalent* of a two-stage stochastic programming formulation:

$$\text{Minimise} \sum_{k \in V_h} g_j y_j + \sum_{\xi \in \Xi} p_\xi \sum_{i \in V_d} \sum_{j \in V_h} q_{i\xi} c_{ij} x_{ij\xi} \tag{2.85}$$

subject to

$$\sum_{j \in V_h} x_{ij\xi} = 1 \qquad \forall i \in V_d, \xi \in \Xi \tag{2.86}$$

$$x_{ij\xi} \leq y_j \qquad \forall i \in V_d, j \in V_h, \xi \in \Xi \tag{2.87}$$

$$x_{ij\xi} \geq 0 \qquad \forall i \in V_d, j \in V_h, \xi \in \Xi \tag{2.88}$$

$$y_j \in \{0, 1\} \quad \forall j \in V_h. \tag{2.89}$$

In this model, constraints (2.86) ensure that each customer is assigned to one facility and therefore served in each scenario. Constraints (2.87) link the first- and second-stage decisions; in particular, they ensure that the assignments in each scenario are consistent with the location decisions made at the first level. This means, for example, that if a facility is not located on a node $j \in V_h$, then no customer can be assigned to this node in any scenario. The rest of the constraints model the nonnegativity and binary restrictions on the decision variables.

We will now show how this model can be applied to the instance in Example 2.5.

Example 2.5 (*Continued*)

For this example, the deterministic equivalent formulation is written as follows:

$$\begin{aligned}
\text{Minimise } & 200y_1 + 200y_2 + 200y_3 \\
& + 0.2(90x_{111} + 410x_{121} + 200x_{131}) \\
& + 0.2(410x_{212} + 90x_{222} + 200x_{232}) \\
& + 0.6(90x_{113} + 410x_{123} + 200x_{133} + 410x_{213} \\
& + 90x_{223} + 200x_{233})
\end{aligned}$$

subject to

$$x_{111} + x_{121} + x_{131} = 1$$

$$x_{112} + x_{122} + x_{132} = 1$$

$$x_{113} + x_{123} + x_{133} = 1$$

$$x_{211} + x_{221} + x_{231} = 1$$

$$x_{212} + x_{222} + x_{232} = 1$$

$$x_{213} + x_{223} + x_{233} = 1$$

$$x_{111} \leq y_1$$

$$x_{112} \leq y_1$$

$$x_{113} \leq y_1$$

$$x_{211} \leq y_1$$

$$x_{212} \leq y_1$$

$$x_{213} \leq y_1$$

$$x_{121} \leq y_2$$

$$x_{122} \leq y_2$$

$$x_{123} \leq y_2$$

$$x_{221} \leq y_2$$

$$x_{222} \leq y_2$$

$$x_{223} \leq y_2$$

$$x_{131} \leq y_3$$

$$x_{132} \leq y_3$$

$$x_{133} \leq y_3$$

$$x_{231} \leq y_3$$

$$x_{232} \leq y_3$$

$$x_{233} \leq y_3$$

$$y_1, y_2, y_3 \in \{0, 1\}$$

All x variables ≥ 0.

2.8.2 Recourse actions

The treatment of the stochastic facility location problem presented in the previous section made one fundamental assumption that allowed writing the deterministic equivalent formulation. The assumption was that, for any choice of the first-stage variables, there exists a feasible solution to the second-stage problem. This assumption always holds for the uncapacitated facility location problem, as there are no limits on the number of facilities that can be located or on the number of customers that could be assigned to a single facility. Indeed, we have seen in the instance of Example 2.5 that an optimal assignment exists even when there is only a single facility located.

The situation is different when second-stage decisions are not always feasible for some of the first-stage decisions taken. Consider, for example, the capacitated facility location problem described in Section 2.7.2 where there are limits on the total amount of demands served by each facility, and that there is now an uncertainty on the demands requested by each customer represented by a finite set of scenarios. This variant of the problem can also be cast in a two-stage stochastic programming framework, in the same way as was done for the uncapacitated counterpart, where the first-stage decisions pertain to the location of the facilities. The difference in this case is that, when the first-stage decisions are made, one of the two cases here will arise in the second stage, separately for each scenario:

1. The facilities located in the first stage have enough capacity to meet the total demand for that particular scenario. In this case, the second-stage problem will have at least one feasible solution.
2. The total capacity made available in the first stage will not be sufficient to meet the total demand. In this case, a *recourse* action should be taken to mitigate the effects of insufficient capacity. Simply penalising the amount of demand that has not been met and outsourcing it from a third-party supplier are some recourse actions. In most cases, such recourse actions will incur extra cost.

Here, we present the deterministic equivalent version of a two-stage stochastic programming formulation of the capacitated facility location with stochastic demands, which is an extension of the model presented in the previous section, but this time with an additional variable $\bar{q}_{i\xi}$ that denotes the amount of demand of customer $i \in V_d$ unmet in scenario $\xi \in \Xi$, and a unit penalty Δ_i to account for failing to meet this amount. In this case, the problem can be formulated as follows:

$$\text{Minimise} \sum_{k \in V_h} f_j y_j + \sum_{\xi \in \Xi} p_\xi \sum_{i \in V_d} \sum_{j \in V_h} q_{i\xi} c_{ij} x_{ij\xi} + \sum_{\xi \in \Xi} \sum_{i \in V_d} \Delta_i \bar{q}_{i\xi} \qquad (2.90)$$

subject to

$$\sum_{j \in V_d} q_{i\xi} x_{ij\xi} \leq \pi_j y_j \qquad \forall j \in V_h, \xi \in \Xi \qquad (2.91)$$

$$\bar{q}_{i\xi} = q_{i\xi} \left(1 - \sum_{j \in V_h} x_{ij\xi} \right) \qquad \forall i \in V_d, \xi \in \Xi \qquad (2.92)$$

$$x_{ij\xi} \geq 0 \qquad \qquad \forall i \in V_d, j \in V_h, \xi \in \Xi \qquad (2.93)$$

$$y_j \in \{0, 1\} \qquad \qquad \forall j \in V_h. \qquad (2.94)$$

In this model, constraints (2.91) model the capacity restrictions. The reader will notice that the model does not contain the assignment constraints in their classical form (2.86) but instead includes a set (2.92) which calculates the amount $\bar{q}_{i\xi}$ of unmet demand. Given that the objective function (2.90) aims to minimise the penalties associated with $\bar{q}_{i\xi}$ with a large enough value of Δ_i, these constraints will strive to meet as much demand as possible in order to achieve $\bar{q}_{i\xi} = 0$, but this might not always be guaranteed. It is therefore possible that $\bar{q}_{i\xi} > 0$ in a given solution to this formulation. The solution also depends on the values of Δ_i, $i \in V_d$ and the trade-off between the costs of meeting the demand (including the fixed location and the shipment costs) and those of any penalties.

Example 2.5 (*Continued*)

Using the facility location instance shown in Example 2.5, assume that each facility has a capacity of $\pi_1 = \pi_2 = 10$ and $\pi_3 = 15$ units, and the unit penalties for unmet demands are $\Delta_1 = \Delta_2 = 100$. The deterministic equivalent of the two-stage stochastic programming formulation can be written as follows:

Minimise $200y_1 + 200y_2 + 200y_3$

$$+ 0.2(90x_{111} + 410x_{121} + 200x_{131})$$

$$+ 0.2(410x_{212} + 90x_{222} + 200x_{232})$$

$$+ 0.6(90x_{113} + 410x_{123} + 200x_{133} + 410x_{213}$$

$$+ 90x_{223} + 200x_{233})$$

$$+ 100(\bar{q}_{11} + \bar{q}_{12} + \bar{q}_{13}) + 100(\bar{q}_{21} + \bar{q}_{22} + \bar{q}_{23})$$

subject to

$$10x_{111} \leq 10y_1$$

$$10x_{212} \leq 10y_1$$

$$10x_{113} + 10x_{213} \leq 10y_1$$

$$10x_{121} \leq 10y_2$$

$$10x_{222} \leq 10y_2$$

$$10x_{123} + 10x_{223} \leq 10y_2$$

$$10x_{131} \leq 15y_3$$

$$10x_{232} \leq 15y_3$$

$$10x_{133} + 10x_{233} \leq 15y_3$$

$$\bar{q}_{11} = 10(1 - x_{111} - x_{121} - x_{131})$$

$$\bar{q}_{12} = 0$$

$$\bar{q}_{13} = 10(1 - x_{113} - x_{123} - x_{133})$$

$$\bar{q}_{21} = 0$$

$$\bar{q}_{22} = 10(1 - x_{212} - x_{222} - x_{232})$$

$$\bar{q}_{23} = 10(1 - x_{213} - x_{223} - x_{233})$$

$$y_1, y_2, y_3 \in \{0, 1\}$$

All x variables ≥ 0.

The optimal solution of this formulation has a total expected cost of 544 units, where two facilities, one at node F_1 and the other at F_2, are open, much in the same way as shown in Figure 2.14. In this case, optimal values of variable $\bar{q}_{i\xi}^*$ are equal to 0 for all $i \in V_d$ and $\xi \in \Xi$, indicating that there is no unmet demand. This is explained by the high unit penalty for any unmet demand.

Now assume that the unit penalties are reduced and changed as $\Delta_1 = \Delta_2 = 20$, and that the fixed cost of opening a facility at node F_3 is changed to $f_3 = 100$. In this case, the objective function of this formulation would be read as follows:

Minimise $200y_1 + 200y_2 + 100y_3$

$$+ 0.2(90x_{111} + 410x_{121} + 200x_{131})$$

$$+ 0.2(410x_{212} + 90x_{222} + 200x_{232})$$

$$+ 0.6(90x_{113} + 410x_{123} + 200x_{133} + 410x_{213}$$

$$+ 90x_{223} + 200x_{233})$$

$$+ 20(\bar{q}_{11} + \bar{q}_{12} + \bar{q}_{13}) + 20(\bar{q}_{21} + \bar{q}_{22} + \bar{q}_{23}),$$

where the set of constraints remain unchanged. The optimal value of the resulting solution is 460 units, where only one facility is open at node F_3. An optimal set of assignments is as follows: $x_{131}^* = x_{232}^* = x_{233}^* = 1$ and $x_{133}^* = 0.5$, and all other assignment variables are zero. The value $x_{133}^* = 0.5$ indicates that only 50% of the demand of customer C_1 can be met from facility F_3 in Scenario 3, which, in turn, results in $q_{13}^* = 5$ units of their demand unmet. In the light of the reduced penalties and the more attractive cost for facility F_3, this option is now much cheaper than the cost of the uncapacitated location instance shown in Figure 2.13, which had a total expected cost equal to 544 units.

2.9 Practical application: Hub location for time-sensitive cargo deliveries

An application of a type of a hub location problem is described by Tan and Kara (2007), which arises in locating hubs for cargo or parcel deliveries where any shipment, from the time of leaving the origin, should arrive at the destination within a guaranteed time window. The particular type of operation considered in this study consists of demand centres, from which parcels originate or are destined to. Each demand centre is assigned to an operation centre, which effectively acts as a hub. The study assumes that the single-allocation strategy was adopted to simplify the managerial aspects. At each operation centre, parcels that need shipping to their destinations (outgoing parcels) are sorted, consolidated and loaded onto a vehicle for forwarding to a different operation centre. Similarly, incoming parcels arriving at an operation centre from another centre are unloaded, sorted and forwarded to their ultimate destinations.

Small vehicles are used to ship parcels between demand centres and operation centres. In contrast, bigger and specialised trucks are used for transfers between hubs, not only due to their larger capacity that allows transporting larger volumes, but also because of their speed which can be faster than that of ordinary trucks. One interesting aspect of the problem concerns the time at which a truck can leave an operation centre, which is important as it affects the total delivery time. In particular, a truck cannot leave an operation centre unless all the outgoing parcels have arrived from their point of origin. This is illustrated in Figure 2.15, which shows four hubs denoted by h_1, h_2, h_3 and h_4 corresponding to operation centres, and six demand centres denoted by i_1, i_2, i_3, assigned to hub h_1, and j_1, j_2, j_3, assigned to h_2. Movements between the hubs are made at a discounted rate on the thick arcs. The thinner arcs show the assignments of the demand centres to the hubs. In this example, a truck is used to ship outgoing parcels belonging to i_1, i_2, i_3 from h_1, which will need to wait at h_1 until all the three shipments arrive. Consequently, its departure time will be dependent on the arrival of the latest parcel to this hub. Similarly, a vehicle leaving hub h_2 to deliver the parcels to the destination points j_1, j_2, j_3 will have to wait for the latest arriving parcel from the remaining hubs h_1, h_3 and h_4, regardless of where it is going.

The optimisation problem here is to design a delivery system from scratch by finding a subset of a given set of nodes in the network to assign as hubs (operation centres) and assign the remaining nodes (demand centres) to the

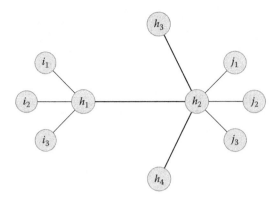

Figure 2.15 The distribution structure used in the hub location application.

hubs. The assumption is that demand centres can be used as hubs, and each operation centre can be a demand centre in itself. As the deliveries are time-sensitive, an upper bound β on time is imposed on the total time for any shipment from its origin to destination. In other words, any shipment, upon leaving the origin at time 0, should arrive at the destination by time β. This is all the more important given that the company measures the service quality by the delivery times. All these considerations give rise to the *Latest Arrival Hub Covering Problem*, the objective of which is to minimise the total number of hubs used.

The application described by Tan and Kara (2007) is for a company already in operation in Turkey with 81 demand points, corresponding to 81 cities in Turkey, and which uses as many as 26 existing operation centres. Solutions are obtained using an integer linear programming formulation of the problem. The discount factor that the company suggested to use is $\alpha = 0.9$. The model is run with various values of β ranging from 36 hours down to 26 hours. In the former case, only one hub is required in an optimal solution. In the latter case, seven hubs are required but this can increase up to nine hubs if the discount factor is reduced to $\alpha = 0.8$. The authors note that the computational time requirements in solving the model to optimality increase substantially when β is reduced for a fixed value of α.

References and further reading

Earlier formulations of hub location problems, including *p*-hub median, uncapacitated hub location, *p*-hub centre and hub covering appear in O'Kelly (1987), Campbell (1994) and Skorin-Kapov et al. (1996). More recent formulations and algorithms are described in Contreras et al. (2011a, 2012) for large-scale uncapacitated hub location problems and in Contreras et al. (2011b) for the single-assignment capacitated hub location problem.

Formulations for the multiple-allocation hub location problem can be found in Marín et al. (2006). Congestion in a hub location has been discussed in and studied by Elhedhli and Hu (2005) and de Camargo et al. (2009). For further reading on the topic, one may consult the surveys by Alumur and Kara (2008), Campbell and O'Kelly (2012), Farahani et al. (2013) and the book edited by Laporte et al. (2015).

The models presented in this chapter are based on those described in the references above.

Chapter 3

Transportation and service networks

In freight transportation networks, demand is generated by *shippers* who often outsource their shipments to external parties called *carriers*. Sometimes these two players are part of the same entity (e.g. company, organisation), particularly when shippers operate their own transportation fleet. Carriers are concerned with designing the transportation network on which they will operate their fleet so as to ensure that freight is shipped, from origin to destination, at the lowest possible cost and as efficiently as possible. Given the long-term span of decisions associated with network design, the demands are often forecasts produced using aggregated data over a period of time, as opposed to individual shipment requests.

Making better decisions to ensure having the right type of infrastructure is the key to providing an efficient and cost-effective transportation network. These are challenging and long-term decisions to make particularly when decision makers are constrained by exogenous parameters such as a constrained budget, or the limitations of the external environment in which the transportation services will operate. Once the transportation infrastructure is in place, additional decisions will need to be made relating to the type of services to run on the network, and planning the way that they will be made. These medium- to short-term decisions are equally difficult yet important in ensuring an effective use of the resources.

The decisions can be examined according to the classical categorisation of planning, namely strategic (long-term), tactical (medium-term) and operational (short-term) level of planning and management of operations. This chapter will present planning problems that fall within the former two of these categories and look into answering the two main questions from a planning perspective:

1. What is the best physical configuration of the network on which transportation services will be run?
2. Given a physical infrastructure, what services (e.g. type, route, schedule) should be offered for demand to be shipped from origin to destination.

This chapter presents and discusses two types of design problems, one at the strategic level and the other at the tactical level of decision making, and presents formulations for both problems.

3.1 Features of transportation networks

A transportation *network* is generally represented by a set of nodes and links, where the former may represent terminals, demand and origin points, and the latter are the physical lines connecting the nodes characterised by their attributes, such as type and capacity, among others.

Network design decisions are at the strategic level of decision making and concerned with having the right transportation infrastructure for shipping freight. They are also referred to as *system design*, as the decisions are made at a system-wide level. These decisions have a long-term impact and often incur significant investment costs. Network design can mean building a transportation network from scratch, although more often the design process will assume that there exist some components of the network or the design process is for expanding the existing infrastructure than for building a new one.

The term *design* may pertain to individual components of a network, such as

- Number and location of facilities (as discussed in Chapters 1 and 2) such as rail yards, intermodal platforms and container ports
- The handling capacity of each facility through decisions, such as the type and quantity of equipment (e.g. cranes) installed at each facility or the space allocated for handling cargo (e.g. containers)
- The type and capacity of links between each facility

as well as a combination of the components. The objective is generally minimisation of a cost function, where the term cost is used in its most generic sense and can include internal or external costs. The following are different types of internal costs:

- Operational costs of vehicle hire or purchase, maintenance, insurance, depreciation
- En-route costs, such as driver or crew wages, and fuel consumption
- Facility costs, such as those associated with rental or building, storage, insurance, administration and warehouse operations

External costs, on the other hand, are difficult to quantify and are often related to the societal and environmental effects, such as congestion, pollution, noise, accidents and land use, which will be discussed in further detail in Section 6.1.

Example 3.1

In rail transportation, en-route costs are associated with running a train with a specified load over its route. It may be composed of crew costs, fuel costs, engine providing and running costs and delay costs. Facility costs may include costs of inspection, classification and delay at yards. In the simplest case, the yard cost function only depends on the total throughput of the yard.

The general network design problem is concerned with the shipment of a set of commodities, mainly characterized by their origin, destination and amount to be shipped, on a network of facilities located on *node*s, where the commodities will flow over the *link*s of the network. There are two possibilities for when it comes to the availability of links:

1. No link physically exists between a pair of nodes. Part of the design problem is to decide on which links to build, which often comes at a fixed (one-off) cost. An example to this is laying physical rail tracks between rail stations for train services.
2. A link physically exists between a pair of nodes. In this case, part of the design problem entails deciding whether to activate the link for use in the transportation of goods. One example to this concerns introducing a new service between a pair of nodes in the network, such as a new trade route between two sea ports on which a container ship will operate.

The links can be in the form of *arc*s if they are directed (also named one-way or uni-directional) or *edge*s if they are undirected (also named two-way or bi-directional). The *fixed cost* associated with installing capacity on each link may be defined in multiples of a unit capacity or in predefined amounts, and a *variable cost* that is proportional to the amount of flow on each link of the network.

The general problem may further be differentiated with respect to the following attributes:

- The problem may be concerned with the shipment of one commodity, known as the *single-commodity* variant, or several commodities simultaneously, known as *multiple*, or more commonly named as the *multi-commodity* variant. In either case, the network is the common resource for consumption by the commodities.
- The problem may include capacity restrictions on the amount of commodities that can flow across the network. Capacity can be defined either with respect to the links or the nodes of the network and defines the maximum amount of flow that can be shipped on a link or be processed in a node, respectively. If the capacity on the network is finite, then the problem is *capacitated*, otherwise it is *uncapacitated*. Of the

former, the version of the problem where there are capacity constraints on the links is more commonly studied than the one with capacity restrictions on nodes.

The objective of network design is usually to design a network with minimum total cost. There are two main sets of constraints in such problems. The first constraint relates to *flow conservation* which dictates that a given amount of each commodity should be sent from its origin to destination through the network. The second constraint concerns *capacity*, which limits the total amount of flow on each link by the installed capacity or the flow into a facility by the capacity of that node.

3.2 Common notation

The standard network design problem is defined on a graph $G = (V, \mathcal{L})$ where V is the set of nodes corresponding to the facilities on the network, such as a rail yard, a container terminal, a seaport or an airport, pick up or delivery points. The standard problem assumes that the facilities already exist and therefore are not part of the design decisions. The set of potential links is shown by the set $\mathcal{L} \subseteq \{\{i,j\}|i \in V, j \in V, i \neq j\}$, that is, $\mathcal{L}$ does not necessarily include links between all pair of nodes. In rail transportation, for example, the set of links may represent physical rail tracks joining the nodes. In maritime transportation, a link may correspond to a trade route defined between two ports. In the rest of the chapter, we will assume that the links are directed and the graph $G = (V, A)$ is defined with respect to a set of arcs defined as $A \subseteq \{(i,j)|i \in V, j \in V, i \neq j\}$.

Here, we differentiate between three types of nodes, namely those with supply for the commodity (e.g. manufacturing plants) and shown by the set $\{i \in V : d_i < 0\}$, those that generate demand for the commodity (e.g. customers, retailers) and shown by the set $\{i \in V : d_i > 0\}$ and finally those that are used as transshipment nodes (e.g. ports, yards) through which the commodity flows and denoted by the set $\{i \in V : d_i > 0\}$. The problem assumes that supply and demand are balanced, that is $\sum_{i \in V} d_i = 0$. If this condition is not satisfied, the problem could be balanced through introduction of artificial nodes that represent excess supply or demand. There are no capacity restrictions on the links, which implies that an unlimited amount of commodity can flow over any arc, although for practical applications a limit $\sum_{i \in V_D} d_i$ can be imposed.

For each node $i \in V$, we define $V_i^+ = \{j \in V|(i,j) \in A\}$ as the set of arcs leaving node i, or 'outgoing arcs', and similarly $V_i^- = \{j \in V|(j,i) \in A\}$ as the set of arcs entering node i, or 'incoming arcs'.

The unit (variable) cost of routing the commodity over arc $(i,j) \in A$ is shown by c_{ij}, and an arc carrying a positive flow incurs a fixed-cost g_{ij}.

The fixed-cost generally refers to that of running a service on the arc, such as a freight train or cargo ship.

3.3 Multi-commodity network flow and design

In the following sections, we will present the two main variants of the network design problem, one with a single commodity and the other with multiple commodities, and also mention special cases of these two variants.

3.3.1 Single-commodity network design problem

The simplest form of the network design problem involves a single commodity, for which w units of demand is to be sent from its origin node $o \in V$ to its destination node $d \in V \setminus \{o\}$. The *single-commodity, uncapacitated, fixed-charge network design problem* is to find a set of flows by activating a set of arcs on the network, so as to minimise an objective function which consists of fixed and variable costs.

Two decision variables for each arc $(i, j) \in A$ are needed to model the problem, namely a nonnegative *flow* variable x_{ij} representing the flow and a binary *design* variable y_{ij} that is equal to 1 if the arc is activated, and 0 otherwise.

A mixed-integer linear programming formulation of the uncapacitated single-commodity fixed-charge network design is as follows:

$$\text{Minimise} \sum_{(i,j) \in A} g_{ij} y_{ij} + \sum_{(i,j) \in A} c_{ij} x_{ij} \tag{3.1}$$

subject to

$$\sum_{j \in V_i^+} x_{ij} - \sum_{j \in V_i^-} x_{ji} = d_i \qquad \forall i \in V \tag{3.2}$$

$$x_{ij} \leq M y_{ij} \qquad \forall (i, j) \in A \tag{3.3}$$

$$x_{ij} \geq 0 \qquad \forall (i, j) \in A \tag{3.4}$$

$$y_{ij} \in \{0, 1\} \qquad \forall (i, j) \in A, \tag{3.5}$$

where

$$d_i = \begin{cases} w, & \text{if } i = o \\ -w, & \text{if } i = d \\ 0, & \text{otherwise.} \end{cases}$$

Constraints (3.2) ensure conservation of flow in the network. Constraints (3.3) are used to ensure that any arc (i, j) carrying a positive amount of flow $x_{ij} > 0$ forces the corresponding design variable y_{ij} to be equal to 1. Note that the definition of the variable x_{ij} allows for splitting of the commodity. The remaining constraints (3.4) and (3.5) are the nonnegativity and integrality restrictions on the flow and design variables, respectively.

The formulation (3.1) through (3.5) can be strengthened through the use of additional valid inequalities. We present one of such inequalities here, for the problem defined on a directed network. The first inequality is written for any subset $V_S \subset V$ with $\sum_{i \in V_S} d_i > 0$ and is given as follows:

$$\sum_{i \notin V_S, j \in V_S} y_{ij} \geq 1. \tag{3.6}$$

These so-called *dicut* inequalities state that for any subset of nodes with a positive demand, there needs to be at least one arc entering set S.

3.3.2 Shortest path problems

A special case of the uncapacitated single-commodity network design problem arises when there are no fixed costs associated with the arcs, or, alternatively, the design decisions concerning the arcs are not relevant. In this case, the problem reduces to finding a minimum cost flow of the commodity from its origin $o \in V$ to its destination $d \in V \backslash \{o\}$, more commonly known as the *shortest path problem*. The shortest path problems arise in a multitude of settings, ranging from finding directions on a road network (e.g. satellite navigation systems used in passenger cars) to routing automated guided vehicles in a container port.

In this case, the amount of the commodity to be shipped can be normalised to $w = 1$ given that the network is not capacitated. There is a cost c_{ij} for travelling on an arc $(i, j) \in A$, which might refer to the time needed to travel from node i to node j, the actual travel cost incurred or any other performance measure that is of concern. The shortest path problems typically aim at minimising the total cost of the path chosen from origin to destination.

A formulation for the problem is given here, where a continuous variable x_{ij} denotes the amount of flow on arc $(i, j) \in A$.

$$\text{Minimise} \sum_{(i,j) \in A} c_{ij} x_{ij}$$

subject to

$$\sum_{j \in V_i^+} x_{ij} - \sum_{j \in V_i^-} x_{ji} = \begin{cases} 1, & \text{if } i = o \\ -1, & \text{if } i = d \\ 0, & \text{otherwise.} \end{cases} \qquad \forall i \in V$$

$$x_{ij} \geq 0 \qquad\qquad \forall (i, j) \in A.$$

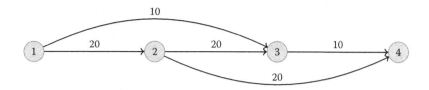

Figure 3.1 The graph for the shortest path problem instance of Example 3.2.

The following example illustrates how this formulation can be applied.

Example 3.2

A parcel is to be shipped from location $o = 1$ to $d = 4$ on the network shown in Figure 3.1. The numbers on the arcs indicate the time it will take for the shipment to be sent over that arc.

This network has only three paths from the origin node to the destination node of the parcel, namely Path 1 following the nodes $(1, 2, 3, 4)$ requiring 50 time units, Path 2 following $(1, 2, 4)$ requiring 40 time units and Path 3 as $(1, 3, 4)$ requiring 20 time units, making the latter the lowest cost (or shortest) path for the parcel to be sent over.

If the aim is to minimise the total time of the journey, the formulation of the corresponding shortest path problem for this instance would read as follows:

$$\text{Minimise } 20x_{12} + 20x_{23} + 10x_{34} + 10x_{13} + 20x_{24}$$

subject to

$$x_{12} + x_{13} = 1 \tag{3.7}$$

$$x_{23} + x_{24} - x_{12} = 0 \tag{3.8}$$

$$x_{34} - x_{13} - x_{23} = 0 \tag{3.9}$$

$$x_{24} + x_{34} = 1 \tag{3.10}$$

$$x_{12}, x_{23}, x_{34}, x_{13}, x_{24} \geq 0. \tag{3.11}$$

While the formulation presented here can be used to solve the shortest path problem using standard off-the-shelf software, there are more efficient methods available that are based on dynamic programming. We will now concern ourselves with two practical extensions of the shortest path problems described in Sections 3.3.2.1 and 3.3.2.2.

3.3.2.1 Resource-constrained shortest path problem

Freight distribution is made using a limited amount of resources. For example, a courier might need to ship a certain product or a commodity from where it originates to where it is destined, at the lowest possible transport cost, but this might be subject to the shipment being delivered within a certain amount of time from leaving the origin node. Some courier companies operate under such guarantees, for example delivery within 24 or 48 hours. Conversely, parcel deliveries that need to be sent at the minimum possible might be subject to a limited budget.

In more formal terms, a shipment that is destined from an origin node $o \in V$ to a destination node $d \in V \backslash \{o\}$ is known to consume h_{ij}^c units of resource when travelling on arc $(i, j) \in A$, where $c \in C$, and C is the index set of the resources needed to carry out shipments (e.g. time, space, cost). There is a total of $\bar{h}^c$ units available for each resource $c \in C$. This additional consideration gives rise to an extension of the shortest path problem, namely the *resource-constrained shortest path problem*, where the aim is to minimise the total cost of the shipment of a single product or a commodity such that resource constraints are not violated. The formulation of this problem is similar to the one shown in Section 3.3.2 but with additional constraints that model resource consumption, as shown here:

$$\text{Minimise} \sum_{(i,j) \in A} c_{ij} x_{ij}$$

subject to

$$\sum_{j \in V_i^+} x_{ij} - \sum_{j \in V_i^-} x_{ji} = \begin{cases} 1, & \text{if } i = o \\ -1, & \text{if } i = d \\ 0, & \text{otherwise.} \end{cases} \qquad \forall i \in V$$

$$\sum_{(i,j) \in A} h_{ij}^c x_{ij} \leq \bar{h}^c \qquad \forall c \in C$$

$$x_{ij} \geq 0 \qquad \forall (i,j) \in A.$$

It is easy to see that if $C = \emptyset$, or if $\bar{h}^c$ is sufficiently large for all $c \in C$, then the formulation is the same as that of the shortest path problem. The resource-constrained variant of the shortest path problem is significantly harder to solve as compared to its unconstrained counterpart.

Example 3.3

Consider the shortest path problem instance given in Example 3.2 and Figure 3.1. If the shipment is to be made under a time constraint of at most 40 units, then set $C = \{c\}$ is a singleton containing a resource c that corresponds to time $h_{ij}^c = c_{ij}$ spent on each route. Setting $\bar{h}^c = 40$

makes path Path 1 infeasible. In this case, it would suffice to add the following constraints to the formulation in Example 3.2:

$$20x_{12} + 20x_{23} + 10x_{34} + 10x_{13} + 20x_{24} \leq 40.$$

Other resources, such as the cost of travel on each arc, can be treated in the same way in case there is limited travel budget.

3.3.2.2 Multi-objective shortest path problem

One well-known extension of the shortest path problem arises when there exist multiple and conflicting objectives that need to be jointly optimised. In network optimisation problems, for example, it is increasingly necessary to take several objectives such as travel cost, travel time, distance, greenhouse gas emissions and transfer times into account. In this case, let $\{1, 2, \ldots\}$ be a finite index set of the objective functions and let $c_{ij} = (c_{ij}^1, c_{ij}^2, \ldots)$ be a generalised cost vector associated with each arc $(i, j) \in A$.

The multi-objective shortest path problem can then be formulated as follows:

$$\text{Minimize } \left(f^1(x), f^2(x), \ldots, \right)$$

subject to

$$\sum_{j \in V_i^+} x_{ij} - \sum_{j \in V_i^-} x_{ji} = \begin{cases} 1, & \text{if } i = o \\ -1, & \text{if } i = d \\ 0, & \text{otherwise.} \end{cases} \quad \forall i \in V$$

$$x_{ij} \geq 0 \qquad \forall (i, j) \in A,$$

where $f^r(x) = \sum_{(i,j) \in A} c_{ij}^r x_{ij}$ with $r = \{1, 2, \ldots\}$.

At this point, it will be useful to introduce some terminology for multi-objective optimisation. Let $x = \{x_{ij}\}$ be a solution vector and let $f(x) = (f^1(x), f^2(x), \ldots)$ be the corresponding vector of the objective function values, which is also called a *point*. For any two solutions x' and x'', $f(x'')$ is said to be a *dominated* point if $f(x'') \leq f(x')$ and $f^r(x'') < f^r(x')$ for at least one $r = 1, 2, \ldots$. If there exists no other point that dominates $f(x')$, then it is said to be *non-dominated* and the corresponding solution x' is said to be *efficient*.

The following example illustrates the application of a special case of the multi-objective shortest path problem with two objectives, alternatively known as the bi-objective shortest path problem:

Example 3.4

Consider the network presented in Figure 3.1 where, in addition to time, each arc has an associated travel cost as shown in Figure 3.2. In this figure, the label (t_{ij}, c_{ij}) appears on each arc, where t_{ij} is the travel time and c_{ij} is the travel cost. This is an instance of the bi-objective shortest path problem. Table 3.1 shows the values of the two objectives for each path.

Let x^i correspond to a solution vector for path i $(i = 1, 2, 3)$. In this case, $f(x^1) = (50, 90)$ is dominated by both $f(x^2) = (40, 50)$ and $f(x^3) = (20, 80)$, which means that the decision maker should not be interested in choosing Path 1 for either objective. The points $f(x^2)$ and $f(x^3)$ do not dominate each other as both x^2 and x^3 are efficient solutions. In this case, the decision maker can choose either Path 2 or Path 3, depending on their preference.

In the example that follows, we show that the number of non-dominated points grows with the number of objectives.

Example 3.5

Consider an incomplete network with 20 nodes with 57 arcs, where each arc (i, j) has up to three objective function cost coefficients c_{ij}^k, for $k = 1, 2, 3$. Table 3.2 shows the input data for the problem instance,

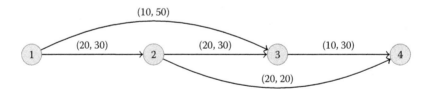

Figure 3.2 Illustration of the shortest path problem with two objectives.

Table 3.1 Three solutions for the bi-objective shortest path network of Figure 3.2

Path	Travel time	Travel cost
(1, 2, 3, 4)	50	90
(1, 2, 4)	40	50
(1, 3, 4)	20	80

Table 3.2   Input data for Example 3.5

Arc (i,j)	c_{ij}^1	c_{ij}^2	c_{ij}^3	Arc (i,j)	c_{ij}^1	c_{ij}^2	c_{ij}^3
(1, 2)	42	68	35	(7, 6)	91	30	46
(2, 3)	35	1	25	(7, 2)	94	49	78
(1, 5)	5	45	63	(10, 4)	85	55	91
(4, 5)	79	59	46	(1, 20)	84	77	56
(4, 14)	20	32	62	(12, 20)	45	10	51
(5, 6)	63	65	43	(7, 4)	38	39	8
(6, 20)	6	46	92	(19, 3)	30	42	88
(7, 8)	82	28	54	(14, 16)	40	59	10
(8, 9)	62	92	22	(5, 11)	7	37	47
(9, 10)	96	43	96	(7, 20)	69	22	1
(1, 11)	28	25	72	(14, 7)	22	46	60
(17, 20)	9	45	13	(1, 4)	43	8	96
(12, 13)	3	54	36	(11, 20)	71	30	22
(3, 14)	93	83	12	(6, 10)	10	59	82
(14, 20)	22	17	74	(2, 9)	23	47	39
(1, 17)	19	96	12	(14, 9)	1	92	29
(16, 20)	48	27	48	(1, 18)	11	36	22
(17, 19)	72	39	58	(5, 18)	49	84	10
(1, 19)	50	13	24	(16, 2)	3	51	50
(19, 20)	16	100	36	(2, 17)	75	21	25
(16, 15)	4	12	41	(17, 2)	49	3	3
(3, 20)	74	35	6	(5, 20)	22	35	24
(2, 12)	5	19	7	(1, 9)	28	68	1
(8, 5)	63	58	24	(18, 8)	54	84	90
(1, 7)	35	32	67	(1, 14)	49	84	23
(18, 20)	24	42	40	(3, 16)	11	60	91
(10, 19)	17	36	19	(2, 20)	89	19	12
(11, 3)	89	7	16	(8, 2)	11	18	29
(1, 3)	65	9	78				

including the arcs that exist in the network and the costs for each of the objectives. Consider now the bi-objective variant of the problem using the costs c_{ij}^1 and c_{ij}^2 shown in the table, and where both objectives involve cost minimisation. In this case, a total of seven non-dominated points exist, which are shown in Figure 3.3. In contrast, the tri-objective variant of the problem instance that uses the three objectives shown under the columns c_{ij}^1, c_{ij}^2 and c_{ij}^3 of Table 3.2 has a total of 11 non-dominated points, which are shown in Figure 3.4.

3.3.3 Multi-commodity network design problem

This section presents a generalisation of the single-commodity, uncapacitated, fixed-charge network design problem that concerns the flow of

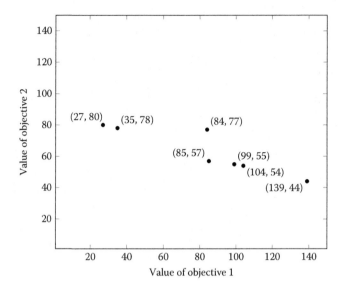

Figure 3.3 The non-dominated set of points for the bi-objective instance of Example 3.5.

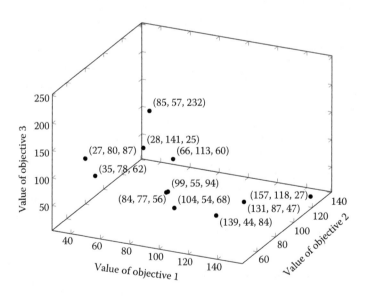

Figure 3.4 The non-dominated set of points for the tri-objective instance of Example 3.5.

multiple commodities denoted by the index set P. Each commodity $p \in P$ has one origin $o(p) \in V$ and one destination $d(p) \in V$, with $o(p) \neq d(p)$, and the quantity w^p that is to be sent from $o(p)$ to $d(p)$. If a commodity has more than one origin or destination, this can be modelled by splitting the commodity into several commodities, each with a single origin and destination. The commodity flows can be translated into demands of nodes as follows:

$$d_i^p = \begin{cases} w^p, & \text{if node } i \text{ is the origin of commodity } p \\ -w^p, & \text{if node } i \text{ is the destination of commodity } p \\ 0, & \text{otherwise.} \end{cases}$$

In this version of the network design problem, each arc $(i, j) \in A$ has a finite capacity that is denoted by u_{ij}, dictated by the limitations on the physical arcs (e.g. maximum number of cars on a given train running over the arc over a given time period) or the mode of transportation operating on the arc (such as the maximum number of containers on a ship or cargo on an airplane). This version is known as the *multi-commodity, capacitated, fixed-charge network design problem*, where the objective is to minimise the total cost of flow and design.

The problem can be formulated as an extension of the single-commodity network design model shown earlier, using nonnegative and continuous flow variables as x_{ij}^p denoting the amount of commodity $p \in P$ carried over arc $(i, j) \in A$. The binary design variables y_{ij} for each $(i, j) \in A$ are as previously defined. The formulation is presented as follows:

$$\text{Minimise} \sum_{(i,j)\in A} g_{ij}y_{ij} + \sum_{(i,j)\in A}\sum_{p\in P} c_{ij}^p x_{ij}^p \tag{3.12}$$

subject to

$$\sum_{j\in V_i^+} x_{ij}^p - \sum_{j\in V_i^-} x_{ji}^p = d_i^p \qquad \forall i \in V, p \in P \tag{3.13}$$

$$x_{ij}^p \leq w^p y_{ij} \qquad \forall (i,j) \in A, p \in P \tag{3.14}$$

$$\sum_{p\in P} x_{ij}^p \leq u_{ij}y_{ij} \qquad \forall (i,j) \in A \tag{3.15}$$

$$x_{ij}^p \geq 0 \qquad \forall (i,j) \in A \tag{3.16}$$

$$y_{ij} \in \{0, 1\} \qquad \forall (i,j) \in A. \tag{3.17}$$

In some applications of the multi-commodity network design problem, the design variables need not be restricted to take binary values and can be allowed to take any integer value, particularly when capacity installation decisions are of concern. The following example presents such an application of the problem.

Example 3.6

Two oil-producing companies are investing in the design of a multi-product oil pipeline network for shared use, which will be used to regularly transport oil between pairs of origin and destination nodes. The first company has a production facility at node 1 in the network shown in Figure 3.5, where the daily rate of production is 100 units, and is to be shipped to node 4 on a daily basis. The production of the second company takes place at node 2, from where 200 units of oil is to be shipped to node 5 on a daily basis. The two oil products are of different types. However, the multi-product pipeline can be used to transport the two products, and any mix that occurs at the interface of the two products (known as *transmix*) is compatible and can be absorbed by either of the two products.

The unit shipping cost for both types of oil is £50 per unit. The set of potential pipelines that can be built is shown by the arcs in Figure 3.5. The capacity of a pipeline is measured as a multiple of a unit level of capacity, for which the values and the corresponding fixed (unit) construction costs are given in Table 3.3. According to the information given in the table, for example, a pipeline installed between node 1 and 2 and carrying flow from the former to the latter node will incur a fixed cost of £500K per 40 tonne capacity, and the cost will increase linearly with any multiple of 40 tonnes.

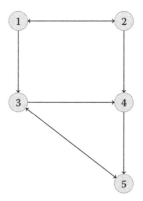

Figure 3.5 Multi-product oil pipeline network of Example 3.6.

Table 3.3 Arc fixed costs and capacities for Example 3.6

Arc	Fixed cost per unit of capacity installed (in £K)	Limit per unit capacity (tonnes)
(1, 2)	500	40
(2, 1)	300	20
(1, 3)	100	30
(2, 4)	100	50
(3, 4)	150	30
(3, 5)	200	10
(4, 5)	300	40
(5, 3)	400	70

An optimal design of the pipeline network can be found by solving the following formulation of the corresponding multi-commodity capacitated network design problem, where the decision variables are defined as follows. Let x_{ij}^k denote the amount of oil flowing on an arc (i, j) of the network shown in Figure 3.5 that belongs to company $k = 1, 2$. We also define an integer variable y_{ij} showing the multiples of the unit capacity installed on each arc (i, j) of the network. For this instance, the formulation takes the following form:

Minimise

$$500{,}000y_{12} + 300{,}000y_{21} + 100{,}000y_{13} + 100{,}000y_{24}$$
$$+ 150{,}000y_{34} + 200{,}000y_{35} + 300{,}000y_{45} + 400{,}000y_{53}$$
$$+ 50\left(x_{12}^1 + x_{12}^2 + x_{21}^1 + x_{21}^2 + x_{13}^1 + x_{13}^2 + x_{24}^1 + x_{24}^2\right)$$
$$+ 50\left(x_{34}^1 + x_{34}^2 + x_{35}^1 + x_{35}^2 + x_{45}^1 + x_{45}^2 + x_{53}^1 + x_{53}^2\right)$$

subject to

$$x_{12}^1 + x_{13}^1 - x_{21}^1 = 100$$
$$x_{24}^1 + x_{21}^1 - x_{12}^1 = 0$$
$$x_{34}^1 + x_{35}^1 - x_{13}^1 - x_{53}^1 = 0$$
$$x_{45}^1 - x_{24}^1 - x_{34}^1 = -100$$
$$x_{53}^1 - x_{35}^1 - x_{45}^1 = 0$$
$$x_{12}^2 + x_{13}^2 - x_{21}^2 = 0$$
$$x_{24}^2 + x_{21}^2 - x_{12}^2 = 200$$

$$x_{34}^2 + x_{35}^2 - x_{13}^2 - x_{53}^2 = 0$$

$$x_{45}^2 - x_{24}^2 - x_{34}^2 = 0$$

$$x_{53}^2 - x_{35}^2 - x_{45}^2 = -200$$

$$x_{12}^1 + x_{12}^2 \le 40y_{12}$$

$$x_{21}^1 + x_{21}^2 \le 20y_{21}$$

$$x_{13}^1 + x_{13}^2 \le 30y_{13}$$

$$x_{24}^1 + x_{24}^2 \le 50y_{24}$$

$$x_{34}^1 + x_{34}^2 \le 30y_{34}$$

$$x_{35}^1 + x_{35}^2 \le 10y_{35}$$

$$x_{45}^1 + x_{45}^2 \le 40y_{45}$$

$$x_{53}^1 + x_{53}^2 \le 70y_{53}$$

$$x_{ij}^k \ge 0 \text{ for all arcs } (i,j) \text{ and } k = 1, 2$$

$$y_{ij} \in \{0, 1, 2, \ldots\} \text{ for all arcs } (i,j).$$

3.4 Service network design

Designing the service network of a consolidation-based carrier involves decisions related to the transportation (or load) plan that will be used to serve the demand in an efficient and a profitable manner. The plans are made on an existing physical infrastructure and a limited amount of resources, as determined during the system design phase described in Section 3.3.

Service network design is concerned with the planning of operations related to the selection, routing and scheduling of services, the consolidation of activities at terminals and the routing of freight of each particular demand through the physical and service network of the carrier. These activities are a part of the tactical planning at a system-wide level and for a given planning horizon, which might be weekly, monthly or yearly. The two main decisions of service network design are to determine the service network and the routing of demand. The former refers to selecting the routes, characterised by origin–destination nodes, intermediate stops and the physical route, and attributes, such as the frequency or the schedule, of each service. The latter is concerned with the itineraries that specify how to move the flow of each demand, including the services and terminals used, the operations performed in these terminals, etc.

Formulations for service network design either assume that the demand does not vary during the planning period (static formulations) or explicitly consider the distribution of demand, as well as the service departures and the movements of services and loads in time (time-dependent formulations). In both cases, the models take some form of the deterministic, fixed-cost, capacitated, multi-commodity network design formulations.

We now present a model for a service network design problem that is modelled on a graph $G = (V, A)$ representing the physical infrastructure of the system and specifies the transportation services that could be offered. The set of services are denoted by the index set S, where each service is characterised by

- *Mode*, which may represent either a specific transportation mode (e.g. rail and truck services may belong to the same service network) or a particular combination of traction and service type
- *Route*, defined as a path in A, from the origin node to the destination node, with intermediate nodes where the service stops and work may be performed
- *Capacity*, which may be measured in load weight or volume, number of containers, number of vehicles (when convoys are used to move several vehicles simultaneously) or a combination thereof
- *Service class* that indicates characteristics such as preferred traffic or restrictions, speed and priority

Each demand is modelled as a product p in a set P of commodities and characterised by its *origin, destination, type of product or vehicle* and *traffic moves* according to itineraries. An itinerary $l \in L_p$ for commodity $p \in P$ specifies the service path used to move part or the whole of demand p, including the origin and destination terminals, the intermediary terminals where operations (e.g. consolidation and transfer) are to be performed, and the sequence of services between each pair of consecutive terminals where work is performed. The demand for product $p \in P$ is denoted by d_p.

Design of a service network includes decisions concerning the services to operate in the transportation plan and the frequency of each service, so as to be able to meet the demand in a given planning period. The objective(s) of the carrier that are to be optimised include one or more of the following:

- Minimisation of total operating cost
- Increasing the quality of service measured by its speed, flexibility and reliability
- Improving service performance measures, which are usually based on delays incurred by freight and vehicles or other predefined performance targets

The following is a small-scale example of a service network design problem.

Example 3.7

Consider a rail network consisting of four rail yards, shown by R_1–R_4, and the rail tracks connecting them as shown in Figure 3.6. There are three different kinds of commodities $P = \{p_1, p_2, p_3\}$ to be transported over this network. In particular, 70 tonnes of commodity p_1 originate from rail yard R_1 everyday, destined for R_4. Similarly, 10 tonnes of commodity p_2 should be sent from origin R_2 to destination R_4 each day. Finally, 10 tonnes of commodity p_3 should be shipped from R_3 to R_4 daily.

There are five potential services that can be operated on this network, indicated by set $S = \{s_1, \ldots, s_5\}$. Let us denote a service s by a triplet (o, d, c), where o and d are the departure and arrival nodes of the service and c is its capacity (in tonnes), and each service runs direct between o and d. The five services for this example are $s_1 = (R_1, R_4, 10)$, $s_2 = (R_1, R_3, 20)$, $s_3 = (R_2, R_3, 5)$, $s_4 = (R_2, R_1, 5)$ and $s_5 = (R_3, R_4, 50)$.

We assume the following available itineraries for the three commodities. It is possible to send part or whole of commodity p_1 using set $L_1 = \{l_1^1, l_2^1\}$ of itineraries, where l_1^1 assumes direct travel from R_1 to R_4 using service s_1. Itinerary l_2^1 starts from R_1 using service s_2, using which commodities are sent to R_3, where they are unloaded from this service and loaded onto service s_5 to be sent to the final destination R_4. As for commodity p_2, there are three different itineraries to choose from as indicated by the set $L_2 = \{l_1^2, l_2^2, l_3^2\}$. Here, l_1^2 follows the route R_2–R_3–R_4 using services s_3 and s_5 successively, l_2^2 travels from R_2 to R_4 via R_1, first using service s_4 followed by s_1, and l_3^2 follows the route R_2–R_1–R_3–R_4, using the combination s_4, s_2, s_5 of services. Finally, there is only one available service for commodity p_3, denoted by the set $L_3 = \{l_1^3\}$, which is a direct travel from R_3 to R_4 using service s_5.

We further assume that service s_1 incurs a daily running cost of £1000 per day per trip. Other services s_2, s_3, s_4 and s_5 cost £1250, £750, £1750 and £150 per day per trip. The unit cost of shipping products p_1, p_2 and p_3 on any arc is £7.50, £3 and £15 per tonne,

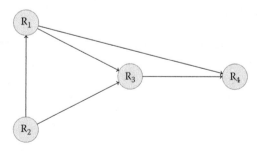

Figure 3.6 The rail network instance of Example 3.7.

respectively, which increases linearly with the amount of product shipped. We consider an objective where the carrier wishes to minimise the total cost of running services and shipping the goods.

3.4.1 Time-invariant formulation

There are at least two types of formulations one can use to solve service network design problems. The first of these is the capacitated network design formulation (3.12) through (3.17) introduced earlier in this chapter, with the caveat that the integer design variables y_{ij} would in this case correspond to the *level* of service offered on arc (i,j), such as frequency of the service during a day or the capacity. This formulation is convenient to use when any sequence of arcs in the network is a possible itinerary for any commodity, which implies that the set of itineraries even for a single commodity is large. Such a formulation is given in Example 3.8.

Example 3.8

A multi-commodity capacitated network design formulation for the service network design problem described in Example 3.7 is as follows. For notational convenience, we use y_{ij} in place of $y_{R_i R_j}$, and similarly x_{ij}^k instead of $x_{R_i R_j}^k$ for a given arc (R_i, R_j) in the graph.

Minimise

$$1250y_{13} + 1000y_{14} + 1750y_{21} + 750y_{23} + 150y_{34}$$
$$+ 7.5\left(x_{13}^1 + x_{14}^1 + x_{21}^1 + x_{23}^1 + x_{34}^1\right)$$
$$+ 3\left(x_{13}^2 + x_{14}^2 + x_{21}^2 + x_{23}^2 + x_{34}^2\right)$$
$$+ 15\left(x_{13}^3 + x_{14}^3 + x_{21}^3 + x_{23}^3 + x_{34}^3\right)$$

subject to

$$x_{13}^1 + x_{14}^1 - x_{21}^1 = 70$$
$$x_{13}^2 + x_{14}^2 - x_{21}^2 = 0$$
$$x_{13}^3 + x_{14}^3 - x_{21}^3 = 0$$
$$x_{21}^1 + x_{23}^1 = 0$$
$$x_{21}^2 + x_{23}^2 = 10$$
$$x_{21}^3 + x_{23}^3 = 0$$

$$x^1_{34} - x^1_{13} - x^1_{23} = 0$$

$$x^2_{34} - x^2_{13} - x^2_{23} = 0$$

$$x^3_{34} - x^3_{13} - x^3_{23} = 10$$

$$-x^1_{14} - x^1_{34} = -70$$

$$-x^2_{14} - x^2_{34} = -10$$

$$-x^3_{14} - x^3_{34} = -10$$

$$x^1_{13} + x^2_{13} + x^3_{13} \leq 20y_{13}$$

$$x^1_{14} + x^2_{14} + x^3_{14} \leq 10y_{14}$$

$$x^1_{21} + x^2_{21} + x^3_{21} \leq 5y_{21}$$

$$x^1_{23} + x^2_{23} + x^3_{23} \leq 5y_{23}$$

$$x^1_{34} + x^2_{34} + x^3_{34} \leq 50y_{34}$$

$$x^k_{ij} \geq 0 \text{ for all arcs } (i,j) \text{ and } k = 1,2,3$$

$$y_{ij} \in \{0,1,2,\ldots\} \text{ for all arcs } (i,j).$$

An optimal solution of this formulation prescribes a service running three times between nodes R_1 and R_3, a single service running between R_1 and R_4 and two services on each of the arcs (R_2, R_3) and (R_3, R_4). In this solution, 60 tonnes of p_1 would be transported from R_1 to R_3, and then to R_4. An additional 10 tonnes would be shipped over the arc (R_1, R_4), making up for the total of 70 tonnes required at the destination. As for commodity p_2, 10 tonnes would be shipped from R_2 to R_4 via R_3. Finally, 10 tonnes of p_3 would be transported on the arc (R_3, R_4). The total cost of this solution is £7735.

A second way of modelling the service network design problem uses what is called a *column generation* formulation, which is preferred for when there are a limited number of services operating on the network and each product has a limited set of itineraries. In this case, it is possible to explicitly represent product–itinerary combinations as decision variables. In particular, the volume of product $p \in P$ shipped using itinerary $l \in L_p$ can be denoted by the continuous variables x^p_l. A second set of integer variables y_s, for each service $s \in S$, is used to represent the frequency with which the corresponding service will run during the planning period.

We will denote the cost of operating service $s \in S$ by $g_s(y)$, assumed to be a function of the frequency of the service. The cost of moving product $p \in P$ by using the itinerary $l \in L_p$ will be denoted by $c^p_l(y, x)$. Finally, a function $Q(y,x)$ will be used to model other types of quality of service or performance measures, which captures various relations and restrictions, such as the limited service or infrastructure capacity. The following formulation can then be used for the service network design problem:

$$\text{Minimize} \sum_{s \in S} g_s(y) + \sum_{p \in P} \sum_{l \in L_p} c_l^p(y,x) + Q(y,x) \qquad (3.18)$$

subject to

$$\sum_{l \in L_p} x_l^p = d_p \quad \forall p \in P \qquad (3.19)$$

$$(y_s, x_l^p) \in X \quad \forall s \in S, l \in L_p, p \in P \qquad (3.20)$$

$$y_s \in \mathbb{Z} \quad \forall s \in S \qquad (3.21)$$

$$x_l^p \geq 0. \quad \forall l \in L_p, p \in P. \qquad (3.22)$$

In this formulation, X are the classical linking constraints akin to (3.14) as well as additional constraints reflecting particular characteristics, requirements and policies of the particular carrier (e.g. particular routing or load-to-service assignment rules). The objective function of this formulation describes a generic cost structure, flexible enough to accommodate various productivity measures related to terminal and transportation operations. As an example, one may consider service capacity restrictions as utilisation targets, which may be allowed to be violated at the expense of additional penalty costs. The last component of the objective function, albeit in a non-linear form, may be used to model such a situation.

We now show how formulation (3.18) through (3.22) can be applied.

Example 3.9

A column generation formulation for the service network design problem instance described in Example 3.7 is as follows:

Minimise

$$1000y_{s_1} + 1250y_{s_2} + 750y_{s_3} + 1750y_{s_4} + 150y_{s_5}$$
$$+ 7.5x_{l_1^1} + 15x_{l_2^1}$$
$$+ 6x_{l_1^2} + 6x_{l_2^2} + 9x_{l_3^2}$$
$$+ 15x_{l_1^3}$$

subject to

$$x_{l_1^1} + x_{l_2^1} = 70$$

$$x_{l_1^2} + x_{l_2^2} + x_{l_3^2} = 10$$

$$x_{l_1^3} = 10$$

$$x_{l_1^1} + x_{l_2^1} \le 10 y_{s_1}$$

$$x_{l_2^1} + x_{l_3^1} \le 20 y_{s_2}$$

$$x_{l_1^2} \le 5 y_{s_3}$$

$$x_{l_2^2} + x_{l_3^2} \le 5 y_{s_4}$$

$$x_{l_2^1} + x_{l_1^2} + x_{l_3^2} + x_{l_1^3} \le 50 y_{s_5}$$

$$y_{s_1}, y_{s_2}, y_{s_3}, y_{s_4}, y_{s_5} \in \{1, 2, \ldots\}$$

$$x_{l_1^1}, x_{l_2^1}, x_{l_1^2}, x_{l_2^2}, x_{l_1^3}, x_{l_3^1} \ge 0.$$

Note the updated objective function coefficients in this formulation, which have been changed to reflect the unit cost of shipping commodities on each itinerary based on the number of arcs that the commodities would be shipped over. For example, itinerary l_2^1 for commodity p_1 involves shipment on two successive arcs which makes for a unit shipping cost of £7.5 × 2 = £15.

An optimal solution to this problem instance indicates that service s_1 should run once a day, service s_2 should run three times a day and services s_3 and s_5 should run twice a day. By doing so, one would be able to transport 10 tonnes of product p_1 on itinerary l_1^1 and 60 tonnes on itinerary l_2^1, which add up to a total of 70 tonnes being shipped from R_1 to R_4. Similarly, 10 tonnes of p_2 should be shipped on itinerary l_1^2 and 10 tonnes of p_3 on itinerary l_1^3. The corresponding service network design would then cost the carrier a total of £7735 per day, where the optimal value for this instance coincides with the value shown in Example 3.8.

3.4.2 Time–space network representation

The service network design model discussed and formulated in Section 3.4.1 was solved for a given planning period, where it was assumed that all products are ready to be shipped at the beginning of the period. The planning period can, for example, be defined as a week or a month, in which the design would be applied on a regular basis (e.g. daily) within this time frame.

However, the representation used earlier does not allow for an explicit representation of the time aspect of the design, particularly in cases where services have different shipment times, and where not all commodities may be available for shipping at the beginning of the period. For instance, in the earlier example, s_2 running from station R_1 to R_3 can be a fast service of one-day duration, whereas s_3 running from R_2 to R_3 can take twice as long. Furthermore, for when there are services which run several times during the planning period, the representation used does not account for the exact point in time at which a particular service departs or arrives. It also cannot accurately model any intermediate storage or idle waiting of the goods, when they arrive at an intermediate node and need to wait to be transferred onto the next available service for the subsequent leg of the itinerary. For example, service s_2 in the network might be running every day of a given week, whereas service s_3 might only have two departures in a week.

When such time considerations are an integral part of the design process, it is convenient to augment the underlying network so as to be able to *implicitly* embed time in the network representation. This necessitates the use of what is known as the *time space* network representation, where the planning period is broken down into individual time units that are appropriate to represent the services and shipments at a suitable level of granularity. The nodes of the network are then replicated for each time unit. If, for example, the planning period is a week, and the minimum travel time that any service requires is a day, then it would be reasonable to construct the time space network based on days. A finer granularity (e.g. hours or minutes) would not provide any benefit and would unnecessarily increase the size of the network in this case. Conversely, if one uses a longer time unit, say two days, then it would be difficult to differentiate between services that, for example, run on a daily basis from those that run every other day. Constructing the time–space network requires that a trade-off be made between the accuracy of the representation of the time aspects and the complexity of the resulting graph. Example 16 shows how a time–space network can be constructed for the instance described in Example 3.7.

Example 3.10

Let us assume that the services discussed in Example 3.7 are to be run in a planning horizon of 5 days, namely Monday to Friday of a given week, and are to repeat themselves every week thereafter. We assume that the services operate according to the information as shown in Table 3.4, where departures take place in the morning at 8am, arrivals are in the evening at 10pm and any loading or transfer of commodities takes place overnight. Based on this example, it is possible to construct a time–space network as shown in Figure 3.7, where the planning period has been divided into 5 days of the week. A copy of each station R_1–R_4 has been made for each day, separately for arrival and departure times.

Table 3.4 Service time information for Example 3.10

Service	Travel time (hours)	Departures
R_1–R_4	14	Only on Tuesdays and Thursdays
R_1–R_3	14	Daily
R_2–R_1	14	Daily
R_2–R_3	38	Only on Mondays and Wednesdays
R_3–R_4	14	Only on Tuesdays, Thursdays and Fridays

The copies are shown using the notation R_i^d, where $i = 1, 2, 3, 4$ is the station number and d is the abbreviation of the day for which the copy has been made, and for which the shorthand notation M = Monday, T = Tuesday, W = Wednesday, Th = Thursday and F = Friday is used. An asterisk next to the day index indicates the 8am departures, whereas an index without an asterisk denotes the 10pm arrivals.

For example, the arc from node R_1^{M*} to R_3^M shows a service which departs from station R_1 on a Monday morning at 8am and arrives at station R_3 on the evening of the same day at 10pm, requiring 14 hours of transit time. Similarly, the arc R_2^{W*} to R_3^{Th} shows a service which departs node R_2 on a Wednesday morning at 8am and arrives into station R_3 at 10pm on the Thursday evening of the same week. The dashed lines between pairs of nodes for a given station show transfers at that node, such as R_1^{M*} to R_1^M.

According to the time–space network representation, the commodity p_1 mentioned in Example 3.7 originating from the yard R_1 and destined to the yard R_4 can use several itineraries, depending on when it is made available. If the commodity is made available to load and ship prior to departures of services on a Monday morning, then it can follow the sequence of nodes R_1^{M*}–R_3^M–R_3^{T*}–R_4^T for arrival at the destination node on the Tuesday evening of the same week, which has a duration of 38 hours, including the 14 hours' transit time between stations R_1 and R_3, 10 hours transfer time at node R_3 and a further 14 hours' travel time on the 8am service from R_3 to R_4. An alternative itinerary for the same commodity with the same duration is R_1^{M*}–R_1^M–R_1^{T*}–R_4^T, which involves an idle wait at station R_1 from Monday 8am until Tuesday 8am. If commodity p_1 is made available, for example on a Tuesday at 1pm, then a possible itinerary is R_1^T–R_1^{W*}–R_1^W–R_1^{Th*}–R_4^{Th}, which will see the arrival of commodity p_1 on the Thursday evening of the same week. This will involve an idle wait at station R_1 from the time of being made available until Thursday 8am, and one day of travel lasting for 14 hours.

Itineraries can be constructed for commodity p_2 originating at the yard R_2 before Monday 8am and destined to the yard R_4. One such itinerary follows the path R_2^{M*}–R_1^M–R_1^{T*}–R_4^T on the time–space network that lasts 38 hours. If the commodity is made available just before 8am on a Wednesday morning, then it can be sent on the path R_2^{W*}–R_3^{Th}–R_3^{F*}–R_4^F, arriving at the yard R_4 on Friday at 10pm.

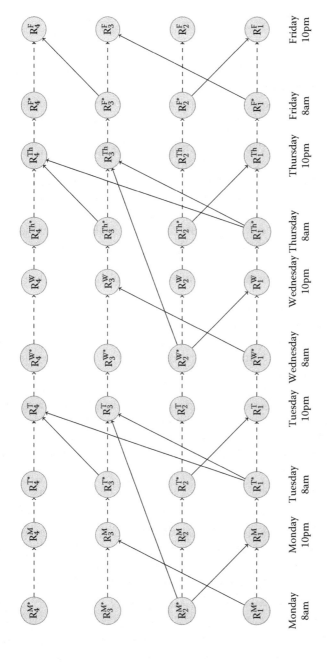

Figure 3.7 The time–space network representation of the services in Example 3.10.

Similar itineraries for commodity p₃ can also be devised. The choice of an itinerary depends on the preferences of the shipper or the logistics provider, which might be based on criteria such as time, cost or environmental indicators.

3.5 Congestion in network design

The significant growth of freight traffic continues to bring significant challenges to the efficient distribution of commodities on transportation networks. One of the more prominent problems is congestion at consolidation hubs, such as ports or rail terminals. Hubs are often the bottlenecks in freight distribution networks. Delays experienced at the hub often propagate through the network and can affect the network performance. Congestion at ports, for example, causes long waiting lines of incoming trucks, resulting in delay and a reduction in productivity. Shippers are less inclined to send their products via congested ports.

Rail transport also receives a fair share of the problem in that rail yards, or even city hubs, can cause bottlenecks in the transportation system. In a rail network, freight cars carrying different types of freight spend most of their time in terminals or rail yards, due to consolidation and classification operations.

The delays experienced in the network can be significant enough to result in breaches in the service level agreements between shippers and carriers, or between the carriers and their customers, in terms of the guaranteed or expected delivery time. Shippers, or more generally *users* of the transportation network, will want to choose not only the cheapest way of transporting their commodities, but also ways in which possible delays along the shipment routes are averted to the best extent possible to achieve a faster transit time. The situation becomes more complex in the presence of multiple users in the network, each with their preferred objective for shipping commodities, particularly when the network resources are limited and will have to be shared.

3.5.1 Modelling delay

There are various approaches to model yard delays, including simulation and analytical models. The latter are primarily based on queueing models and can be easily incorporated in tactical decision models. One of these that can be used to calculate the mean classification delay at a consolidation node is given as follows:

$$
\frac{\Omega\mu}{\Omega - \mu f}, \tag{3.23}
$$

where

    Ω denotes the length of the planning period

    μ is the mean service time at the node

    f is the total amount of traffic flowing into the node

One other analytical expression that is not based on trains but on individual freight cars is given here:

$$\delta + \beta \left(\frac{f}{\Pi} \right)^{\alpha}, \tag{3.24}$$

which is used to calculate the average classification delay for a freight car in a yard. In this expression, δ is the classification delay for a freight car under free flow conditions, f denotes the amount of freight cars to be classified in the yard during the period of analysis, Π is the classification capacity of the yard over the time of analysis and β and α are the calibration parameters. Since this function measures the average classification delay for a freight car in a particular yard, the total delay in the yard with a total flow of f freight cars will be

$$\delta f + \beta \frac{f^{\alpha+1}}{\Pi^{\alpha}}. \tag{3.25}$$

Expressions of type (3.25) are based on the delay functions initially proposed by the U.S. Bureau of Public Roads.

3.5.2 Network design with node congestion

Node capacity is a notion that is often defined with respect to a length of time. In classification yards where incoming freight trains are separated into blocks of cars, and these blocks are used to form new trains which leave the yard for their next destination, capacity can be defined as the amount of cars per unit of time. The MacMillan yard, the largest classification yard in Canada, for example, operates 24 hours a day and handles over a million cars per year. Ports have similar units of measurement in capacity, generally measured by a standard container, which is a 20-foot box, 20 feet long, $8'6''$-feet high and 8 ft wide, referred to as a Twenty-Foot Equivalent Unit (TEU). The Trinity Terminal in Felixstowe, UK, as the largest container handling facility in the United Kingdom for example, has an annual capacity to handle four million TEUs. Larger ports, such as the Ports of Singapore and Shanghai, have handled well over 35 million TEUs each year in the past. Such ports also see regular increases in capacity, be it through expanding the infrastructure (e.g. additional warehousing facilities or berths), or increasing efficiency through better planning. Figure 3.8 shows the way in which

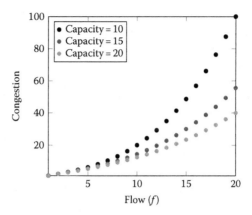

Figure 3.8 Behaviour of congestion function (3.25) at different capacity levels.

function (3.25) behaves for varying amounts of flow $1 \leq f \leq 20$ into a node for three levels of capacity, using the following set of parameters: $\alpha = 2$, $\beta = 1$ and $\delta = 1$. As the figure indicates, increasing the capacity alleviates the effects of congestion (e.g. delay) which is assumed to be proportional to the values shown in the vertical axis. It is important to bear in mind that one will have to investigate the trade-off between investment into additional capacity and the benefits this brings with reduced congestion.

One way to look into the trade-off between increase in capacity and reduction in congestion-related effects in network design problems is to explicitly incorporate analytical models of congestion into the relevant formulations. We will show how this can be done for the particular case of modelling congestion at nodes. The problem is an extension of the capacitated multi-commodity network design problem presented in Section 3.3.3, which we refer to as *congested network design*. The problem explicitly captures congestion at nodes which have limited capacities and allows for increasing node capacities in a one-step fashion to be able to reduce any congestion-related cost when needed. In particular, the initial capacity $\Pi_i > 0$ at each node $i \in V$ can be increased by an amount $\tilde{\Pi}_i > 0$ but at a fixed cost $\hat{g}_i \geq 0$ (but often $\hat{g}_i > 0$). The one-level capacity upgrade assumed here can be found in practice, especially in rail yards. For example, the MacMillan Yard in Toronto, Canada, has a dual hump with two tracks, one of which is often inactive. The existence of the dual lead tracks, and the possibility to activate the second track when needed, provides for the one-step capacity expansion.

To formulate this problem, we use the following parameters and variables, in addition to those introduced in Section 3.3.3. In addition, a new binary variable z_i is equal to 1 if node $i \in V$ is upgraded, and equal to 0 if not.

The free-flow congestion at yard $i \in V$ is denoted by δ_i, and the unit congestion cost is shown by σ_i.

The function we use to measure congestion is (3.25) mentioned earlier, which, using the problem variables and parameters, takes the following form:

$$G_i(x, z) = \sigma_i \left(\delta_i \sum_{j \in V^-} \sum_{p \in \mathcal{P}} x_{ji}^p + \beta \frac{\left(\sum_{j \in V^-} \sum_{p \in \mathcal{P}} x_{ji}^p \right)^{\alpha+1}}{(\Pi_i + \tilde{\Pi}_i z_i)^\alpha} \right). \tag{3.26}$$

A formulation for the problem is given as follows:

$$\text{Minimize} \sum_{(i,j) \in A} g_{ij} y_{ij} + \sum_{(i,j) \in A} \sum_{p \in \mathcal{P}} c_{ij}^p x_{ij}^p + \sum_{i \in V} \hat{g}_i z_i + \sum_{i \in V} G_i(x, z) \tag{3.27}$$

subject to

$$\sum_{j \in V_i^+} x_{ij}^p - \sum_{j \in V_i^-} x_{ji}^p = d_i^p \qquad \forall i \in V, p \in \mathcal{P} \tag{3.28}$$

$$x_{ij}^p \le w^p y_{ij} \qquad \forall (i,j) \in A, p \in \mathcal{P} \tag{3.29}$$

$$\sum_{p \in \mathcal{P}} x_{ij}^p \le u_{ij} y_{ij} \qquad \forall (i,j) \in A, \tag{3.30}$$

$$\sum_{j \in V^-} \sum_{p \in \mathcal{P}} x_{ji}^p \le \Pi_i + \tilde{\Pi}_i z_i \quad \forall i \in V, \tag{3.31}$$

$$y_{ij} \in \{0, 1\} \qquad \forall (i,j) \in A \tag{3.32}$$

$$x_{ij}^p \ge 0 \qquad \forall (i,j) \in A, p \in \mathcal{P} \tag{3.33}$$

$$z_i \in \{0, 1\} \qquad \forall i \in V. \tag{3.34}$$

The objective function represents the total cost of design, routing and capacity augmentation and cost of congestion. In this formulation, (3.28) are the flow conservation constraints which ensure that the demands are satisfied for each node. Constraints (3.29) make sure that the flow of any commodity on an arc is zero when that arc is not selected. Constraints (3.30) imply that the amount of flow on an arc can be at most equal to the capacity of the arc. Finally, constraints (3.31) limit the total inflow of node i either by its initial (i.e. $z_i = 0$) or extended (i.e. $z_i = 1$) capacity. Integrality and nonnegativity restrictions on the decision variables are given by (3.32) through (3.34).

In the following example, we show the effect of incorporating congestion into network design, both on the resulting network configuration and on various cost components.

Example 3.11

A third-party logistics provider wishes to design a distribution network over a set of 10 nodes to regularly ship 10 different types of commodities, each with different sources, destinations and volume. Tables 3.5 and 3.6 show the detailed commodity and node data relevant to the problem instance. There are a total of 35 possible arcs in the network, for which the relevant data including the fixed costs, variable costs (assumed to be the same for all commodities, i.e. $c_{ij}^p = c_{ij}$) and the capacities are shown in Table 3.7. Finally, for each node in the network, the relevant data including the initial capacity, the additional capacity (which is assumed to be 30% of the initial capacity of a node), the upgrade cost and the unit congestion cost are shown. In calculating the

Table 3.5 Commodity data for Example 3.11

Commodity p	Origin node o(p)	Destination node d(p)	Volume w^p
1	8	6	71
2	6	10	32
3	4	2	78
4	4	7	40
5	2	8	53
6	10	4	80
7	3	1	57
8	8	1	72
9	6	1	57
10	9	8	73

Table 3.6 Node data for Example 3.11

Node i	Initial cap. Π_i	Incremental cap. $\tilde{\Pi}_i$	Upgrade cost e_i	Congestion cost σ_i
1	320	256	3294	96.09
2	275	220	1751	59.31
3	229	183	2329	95.11
4	97	77	189	18.28
5	86	69	165	18.00
6	269	216	1503	52.06
7	135	108	850	58.86
8	248	198	2302	86.72
9	86	69	978	106.88
10	107	86	504	43.92

Table 3.7 Network data for Example 3.11

Arc (i,j)	Variable cost c_{ij}	Capacity u_{ij}	Fixed cost g_{ij}
(1,2)	100	613	595
(1,10)	54	81	244
(1,8)	45	151	375
(1,7)	18	223	369
(2,3)	100	613	595
(2,4)	13	198	441
(2,7)	92	115	340
(2,9)	13	215	179
(2,8)	93	193	363
(3,4)	100	613	595
(3,2)	27	265	554
(3,1)	76	129	143
(3,7)	11	143	238
(3,10)	25	129	518
(4,5)	100	613	595
(4,3)	95	232	108
(4,10)	50	179	322
(5,6)	100	613	595
(5,2)	34	73	125
(5,8)	72	232	536
(5,3)	59	276	524
(6,7)	100	613	595
(6,10)	41	101	304
(6,8)	73	173	393
(7,8)	100	613	595
(7,1)	96	198	286
(7,9)	22	221	572
(8,9)	100	613	595
(8,3)	41	134	340
(8,7)	78	134	357
(8,2)	30	201	256
(9,10)	100	613	595
(9,6)	50	240	298
(10,1)	100	613	595
(10,6)	48	156	375

congestion using the function (3.26), it is assumed that $\delta_i = 1$ for all $i = 1, \ldots, 10$, with $\alpha = 3$ and $\beta = 1$.

Figure 3.9 shows the solution to the (standard) multi-commodity network design problem on this instance where node capacities and effects of congestion are not taken into account. The values of flows on each arc given in Table 3.8, which indicate a total flow cost of 69,492 units, which, combined with the total fixed cost of using the arcs, results in a total operational cost equal to 74,079 units. In this solution, the implicit cost of congestion which is not taken into account in solving the problem is calculated as 237,507.18 units, which, combined

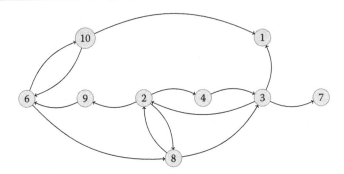

Figure 3.9 Solution of the network design instance in Example 3.11 without congestion.

with the operational cost, results in an overall cost for this solution at 311,586.18.

In contrast, if congestion is explicitly accounted for in the design of the network, then a solution to the corresponding congested network design problem is shown in Figure 3.10. The flow values associated with this solution are given in Table 3.8. Here, the total flow cost is equal to 77,798, which, when combined with the fixed cost of the arcs, yields a total operational cost equal to 83,592 units. Note that this cost is higher than that obtained with the previous solution. In addition, there is an additional cost equal to 3711 units resulting from upgrading nodes 4, 5, 6, 8, 9 and 10, as shown in Figure 3.10. The new solution has a congestion cost equal to 131,009.51 units, resulting in an overall cost equal to 218,312.51 units for this instance, which is a reduction of approximately 30% over the previous solution.

3.6 Practical application: Intermodal rail network service design

This case study is based on Andersen and Christiansen (2009) and Bauer et al. (2010) which relates to the service network design of an intermodal rail network operating over Austria, the Czech Republic and Poland, named Polcorridor, intended to connect northern and southern Europe for shipping cargo between the two regions. The rail network is a part of this corridor, located between two Polish hubs, one in Gdansk and the other in Szczecin, and the main inland hub in Vienna. External services are used to ship freight from or to the two hubs in Poland, namely Swinoujscie (connected to Szczecin) and Gdynia (connected to Gdansk). Similarly, external services from or to Vienna exist to destinations located in southern Europe.

There are three other hubs in the network, one located in Wroclaw, the second in Miedzylesie/Lichkov and the third in Chalupki/Bohumin.

Table 3.8 Solutions for Example 3.11

Arc (i, j)	Without congestion		With congestion	
	Amount of flow	Flow cost	Amount of flow	Flow cost
(1, 2)	0	0	0	0
(1, 10)	0	0	0	0
(1, 8)	0	0	0	0
(1, 7)	0	0	0	0
(2, 3)	0	0	0	0
(2, 4)	80	1,040	80	1,040
(2, 7)	0	0	0	0
(2, 9)	71	923	0	0
(2, 8)	53	4,929	52	4,836
(3, 4)	0	0	0	0
(3, 2)	78	2,106	0	0
(3, 1)	129	9,804	56	4,256
(3, 7)	40	440	39	429
(3, 10)	0	0	0	0
(4, 5)	0	0	72	7,200
(4, 3)	118	11,210	39	3,705
(4, 10)	0	0	4	200
(5, 6)	0	0	0	0
(5, 2)	0	0	72	2,448
(5, 8)	0	0	0	0
(5, 3)	0	0	0	0
(6, 7)	0	0	0	0
(6, 10)	89	3,649	87	3,567
(6, 8)	153	11,169	157	11,461
(7, 8)	0	0	0	0
(7, 1)	0	0	71	6,816
(7, 9)	0	0	0	0
(8, 9)	0	0	70	7,000
(8, 3)	72	2,952	0	0
(8, 7)	0	0	71	5,538
(8, 2)	151	4,530	84	2,520
(9, 10)	0	0	0	0
(9, 6)	144	7,200	143	7,150
(10, 1)	57	5,700	56	5,600
(10, 6)	80	3,840	84	4,032
Total		69,492		77,798

The latter two require that freight has to cross the border between Poland and the Czech Republic. Given the differences in the services, signalling systems and operators between the two countries, freight will need to be transferred from one service to another at these nodes, for which reason the rail services on either side of the border are treated as different modes from a freight planning point of view. However, this is not the case for

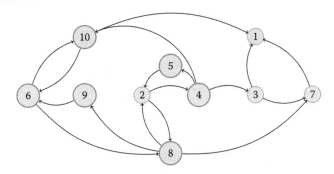

Figure 3.10 Solution of the network design instance in Example 3.11 with congestion, where the upgraded nodes are shown by the larger circles.

services crossing the Czech/Austrian border as the rail systems are compatible between the two countries. Services from Swinoujscie can travel to Wroclaw through Zielona Gora or Poznan. The Polcorridor network in its abstract form is given in Figure 3.11. As the physical network is in place, the planning problem in this case study is to design services that will operate regularly, using weekly schedules. To construct a time–space network representation, a 2-hour time interval is used, which is the minimum time unit needed for any operation in the network (e.g. for travel or for intermodal transfer). This results in 84 time periods to model a week on the basis of 24-hour days. There are 40 commodities that become available at a given point in time in a week, each with a given origin to a given destination. For example, a commodity is brought to Vienna at time 81 in a given week and needs to be shipped to Gdynia as quickly as possible. We assume that it needs 4 hours for intermodal transfers, either across the borders or between the external nodes and the rail network. If the service network design is done in such a way that there are three services operating from Vienna to Bohumin and four services from Chapulki to Gdynia, then a possible itinerary for this commodity is given in the first half of Table 3.9 under *Fast services*, where the operations are denoted by W = wait, IT/E = Intermodal Transfer between External services, IT/CB = Intermodal Transfer over Cross Border and T = Transport. The table also shows when each operation starts and ends, and for transport services, the frequency with which they run over the week in the last column.

According to Table 3.9, after the commodity becomes available at time 81 in a given week, it is kept waiting at Vienna until time 7 of the following week, denoted by '(+1)', after which it is transferred onto the rail network from the external services and then shipped to Bohumin at time 9. The commodity arrives in Bohumin at time 19, transferred over the border for which two time units (equal to 4 hours) is needed. Following this

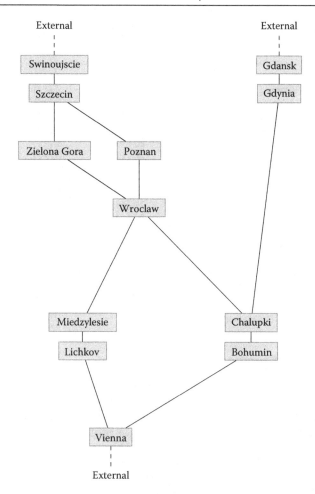

Figure 3.11 Graphical representation of the Polcorridor network.

operation, it is shipped to Gdynia at time 20, arrives there at time 28 and transferred to external services, ready to be fed to the northern network at time 30.

Alternatively, if the frequency of the service from Vienna to Bohumin is reduced to once a week, then the itinerary of the same commodity changes to what is shown in the lower half of Table 3.9 under *slow services*, which follows the same order of stations as with fast services but with different waiting times. In particular, it is kept waiting at Vienna for much longer and subsequently arrives in Gdynia at time unit 12 in the second week of it becoming available, denoted by (+2).

Table 3.9 Two alternative itineraries for a commodity on the Polcorridor rail network

From	To	Operation	Time start	Time end	Frequency
Fast services					
Vienna	Vienna	W	81	7 (+1)	
Vienna	Vienna	IT/E	7 (+1)	9 (+1)	
Vienna	Bohumin	T	9 (+1)	19 (+1)	3
Bohumin	Chapulki	IT/CB	19 (+1)	20 (+1)	
Chapulki	Gdynia	T	20 (+1)	28 (+1)	4
Gdynia	Gdynia	IT/E	28 (+1)	30 (+1)	
Slow services					
Vienna	Vienna	W	81	24 (+1)	
Vienna	Vienna	IT/E	24 (+1)	26 (+1)	
Vienna	Bohumin	T	26 (+1)	36 (+1)	1
Bohumin	Chapulki	IT/CB	36 (+1)	37 (+1)	
Chapulki	Chapulki	W	37 (+1)	2 (+2)	
Chapulki	Gdynia	T	2 (+2)	10 (+2)	4
Gdynia	Gdynia	IT/E	10 (+2)	12 (+2)	

Source: Reprinted from *Journal of the Operational Research Society*, Minimizing greenhouse gas emissions in intermodal freight transport: An application to rail service design, 61, 2010, pages 530–542, J. Bauer, T. Bektaş, T.G. Crainic, with permission of Springer.

References and further reading

An excellent introduction to network design problems that arise in transportation planning is given by Magnanti and Wong (1984), which includes models, special cases and optimisation algorithms. For an introduction to capacitated multi-commodity network design problems, see Gendron et al. (1999).

Service network design problems arising in freight transportation have been discussed in detail by Crainic (2000), and applications to rail freight transportation appear in Crainic and Rousseau (1986) and Zhu et al. (2014).

Earlier references that discuss congestion in freight transportation include Petersen (1977) and Crainic et al. (1984), with a particular focus on rail yards and rail transportation. Modelling and solving network design problems with node congestion have been described in, for example Paraskevopoulos et al. (2016) and Bektaş et al. (2010).

The models presented in this chapter are based on those described in the references above.

Chapter 4

Routing problems

Short-haul transportation is a main component of distribution and logistics. In this context, there exist numerous planning problems arising in the assignment, routing and scheduling of vehicles. This chapter discusses two important problems in this context: the travelling salesman problem (TSP) and the vehicle routing problem (VRP), both specifically aimed at short-haul freight transportation activities.

4.1 Routing in freight distribution

The main goal of freight transportation is to serve, which might refer to *delivery* or *collection*, one or multiple types of goods to or from a given set of origin nodes, each of which is called a *depot*. Service is provided to a set of destinations, each of which is called a *demand point* or a *customer*, using one or many vehicles. Freight distribution, in general, can be done in one of the two following ways, or sometimes using a combination of the two:

1. *Full truckload* (FTL): This arises when the amount of freight requested from the origin by a customer is enough to fill the capacity of the vehicle, also referred to as a *direct shipment* or *direct distribution*. In this case, shipments are direct from an origin node to a destination node.
2. *Less-than full truckload* (LTL): When smaller amounts of one or several types of goods are requested (either for collection or delivery) by several customers, it may be more effective to consolidate the orders into as few vehicles as possible, in which case each vehicle will have to follow a route in which multiple stops will be performed. In case of deliveries only, the total amount of goods carried on a vehicle will be limited by the capacity of the vehicle, which can be expressed in terms of payload, volume or other restrictions.

LTL operations give rise to a wide range of planning problems, arising from the need to consolidate goods. These problems are generally known as vehicle routing and scheduling and are defined with respect to various objectives and under a set of constraints. Such problems often appear in the domain or road transportation and are relevant to an operational level of planning. However, it is not unusual to see similar problems arising in other modes of transport such as maritime. The rest of the chapter will focus on LTL operations and the associated problems arising in the planning of such operations.

First, we discuss a number of key aspects of the environment in which such problems arise.

4.1.1 Network topology

Freight distribution is assumed to be performed on a network, including nodes that correspond to origins and destinations. An origin node is often called a *depot*, where goods to be distributed are assumed to be available, where vehicles are loaded or unloaded and where vehicles start their routes from and return to. The minimum number of depots is one, which is usually the case for most routing problems. When there are multiple depots, additional decisions need to be taken. For example, one needs to decide whether a vehicle returns to the depot where it has started its tour (so called the *fixed-destination* case), or whether it is allowed to return to another depot (which is the *non-fixed-destination* variant). Other decisions relevant to the nodes involve the possibility of transhipments between depots, and the opening hours.

The links between the nodes can be directed (arcs) or undirected (edges). If the network is directed, the cost of traversing a certain arc depends on the direction in which the arc is traversed. In this case, the costs are said to exhibit an *asymmetric* pattern. If the network is undirected, the edge travel cost is independent of the direction travelled and only depends on the pair of nodes connected by the link, in which case the network is said to be *symmetric*. The use of both arcs and edges in the same network is possible.

4.1.2 Timescales for planning and implementation

Freight distribution can be planned at three levels, including strategic, tactical and operational. Some problems arising in the former two were discussed in Chapter 3. Operational plans, on the other hand, are made to be implemented over a relatively short time span. Most LTL operations, for example, are carried out daily, where the devised routing plans are finalised either the day before or early on the same day. Weekly plans can sometimes be made for when the distribution takes place over longer distances. We will refer to the interval of time in which the plans are made and executed as the *planning horizon* or the *planning period*.

In some cases, the planning horizon is broken down into smaller implementation periods, in which the plans are operationalised. For example, provided that there is sufficient information available on the input data, then a plan for an LTL distribution activity might entail the routing plans for every day of a given week, to be made and finalised at the start of the week.

4.1.3 Customers

Customers are located on the nodes in the network. In many cases, the customers, their locations and other relevant information (e.g. demand) are known upfront and the routing plans are devised for these customers. This is not always the case in practice, however, as sometimes customer locations are revealed gradually, even during the execution of the distribution operations, leading to dynamic or real-time variants of the VRP.

Customers will either request a collection (pick up) or delivery of a particular good, and the demand is the amount of pick up or delivery requested. The information on the customer demand is usually known in advance and in a *deterministic* way. This information may pertain to the size or the amount of demand, or even whether there will be a demand at all at a particular node. In other words, it is assumed that one has the actual demand information prior to planning the distribution operations, and this information is not subject to any change. When there is uncertainty in the demand information, then the input data is treated as *stochastic*, which means the associated information (e.g. size) can be represented using a random variable with a known distribution. In this case, demand is revealed at the moment the customer is visited. This could potentially lead to issues with the vehicle capacity or with the route feasibility. For example, if the amount of (stochastic) demand revealed at a customer node is much larger than was planned for, and if the vehicle capacity limits are violated, then an alternative course of action is needed, often referred to as *recourse*, to overcome the difficulties and to restore the feasibility.

Each customer is often restricted to be visited by a single vehicle in a given distribution or routing plan, an assumption which is based on realistic considerations. One consideration is the convenience of the customer itself, for whom it will be easier to be served by a single driver and only once during the implementation time unit (e.g. day). It is less convenient for a customer to receive multiple deliveries, for example, as this implies multiple handling operations, and potentially dealing with different drivers. Another consideration is the potential gain in efficiency for a single driver dealing with a given customer, particularly if a driver is already accustomed to the route of getting to the customer location. In the latter case, there might be additional restrictions that the *same* driver visit a given customer at each implementation period, giving rise to *consistency* constraints. An alternative way of visiting the customers is to employ a strategy known as *load splitting* or *split deliveries*, which arises when a vehicle is allowed to partially serve a customer request. Assuming a delivery operation, for example load splitting, requires that the customer should receive the total amount of goods requested by the end of the planning period, but it will be possible to do this by using several vehicles, visiting the customer at different points in time and delivering partial amounts.

4.1.4 Time

There are three different aspects of time that form an integral part of distribution or routing problems, which are discussed as follows:

1. *Travel time* refers to the time required to traverse a given link in the network, often a function of distance and speed at which a vehicle travels on the link, and is affected by, among other things, traffic conditions and congestion. The travel time for a given link can be time-dependent, meaning that it can change over time. For example, the average speed with which a link is traversed is likely to be much slower during morning or evening rush-hours, as opposed to other times of a day, which would imply a longer travel time for this link during such periods. Alternatively, the travel times might be time-independent, indicating that the time required to travel over a link is constant and does not change with the time of day at which the link is traversed. Travel times can also be deterministic or stochastic, similar to the discussions in customer demand.
2. *Time windows* are intervals of time in which the customer wishes the service to commence and are characterised by the *earliest start* and *latest end* time. Sometimes the request is strict in that it is not possible to serve a customer before the earliest start and after the latest end time, in which case the problem is said to have *hard time windows*. This is the case when the customer is physically not present at the corresponding location, either unable to receive deliveries or does not have the goods ready to be collected. For example, a convenience store might request that the delivery of a fresh product, such as milk, is done early in the morning (e.g. between 7 and 9am). A charity shop to where donations are made might need for the items to be picked up after the store is closed but before the employees leave work (e.g., between 5 and 6pm). Sometimes, parking restrictions on roads near where customers are located necessitate loading and unloading operations to be carried out only at certain times of a day.

   Under hard time windows, if a vehicle arrives at a customer before the earliest start time of the time window, then it is permitted to wait until the earliest start time. The service can then commence after the earliest start time, although the idle wait might incur additional cost, including those associated with increased driver time and parking. Hard time windows do not allow arrivals to customers after the latest end time; if this is the case, then the problem is infeasible. Alternatively, some customers might allow for violations of the early start and latest end times, in which case the problem is said to have *soft time windows* and where there will be penalties associated with per unit of time of early or late arrivals. A solution of a routing problem which satisfies hard time window constraints is also feasible with respect to the case if the same time windows were soft, but the converse is not necessarily true.

3. *Service time* is the length of time that is spent at a given customer location for the service to be completed and might relate to loading or unloading of vehicles, completion of the associated paperwork, or any payments or transactions to be made. Service times are often assumed to be independent of the type of service or the amount to be delivered or collected, although there are cases when they need to be specified of the function of the two. Similar to travel times, service times can be treated as being deterministic or stochastic. The latter might arise particularly in the case of stochastic demands, where the amount of delivery or collection is unknown, which would subsequently affect the service time. With stochastic service times, it is assumed that the real service time will not be revealed until service commences.

4.1.5 Vehicles

The fleet of vehicles used in the distribution operations is an important aspect of the plans to be devised. Vehicles in this context include any means, motorised or otherwise, by which distribution is performed but will most often refer to trucks, lorries or vans in road transport, railway wagon or carriage in rail transport, cargo or container ships or barges in maritime transport or aircrafts in air transport. If all vehicles in a given fleet are of the same configuration, or they are the same *type* of vehicle, then the fleet is said to be *homogeneous*. Otherwise, the fleet is said to be *heterogeneous*. The most commonly used criterion with respect to determining homogeneity is vehicle capacity. For instance, the physical capacity of a vehicle (e.g. volume, weight) will limit the type and amount of goods that can be carried. In some applications, vehicle capacity imposes no limitations, as might be in postal distribution where the total amount of letters or parcels carried is little compared to the available capacity.

Another aspect of the fleet is the size, which refers to the number of vehicles available within the fleet. In some applications, fleet size will be limited in applications such as distribution companies that own their own fleet.

In other applications, an unlimited number of vehicles are assumed to be available, as might be the case where vehicles are available for hire, but this will often come with additional cost associated with hiring a vehicle per unit of time (e.g. hour or day).

4.1.6 Objectives

The set of objectives is the key mechanism through which solutions can be evaluated, with respect to given criteria, and often involves one or more of the following:

- To minimise the total *travel time* of all the vehicles as this impacts the driving times and the corresponding driver costs
- To minimise the *number of vehicles* used to perform the distribution operations, which, in turn, will minimise the associated costs of purchasing or hiring the fleet and any maintenance costs
- To minimise a *cost* function, which can include actual cost incurred in relation to distance, driver times, size and mix of vehicles in the fleet, fuel consumption, loading or handling, or a combination thereof
- To minimise any *penalty* for late visits to customers
- To maximise a *profit* function, expressed as the difference between any revenues obtained from visits to customers (e.g. sales of a particular product) less the costs of performing such visits
- To minimise any *risks* or *hazards* associated with shipments

Depending on the type of application, either a single objective or combinations of several objectives can be considered. In the latter case, multiple objectives can be converted into a single objective using various weights. Alternatively, they might be treated within the framework of multi-objective optimisation techniques.

The concepts described in the previous section along with the various limitations and objectives give rise to interesting and challenging vehicle routing problems. In this section, we will present a number of fundamental and well-known problems that arise at the operational level of planning. In particular, we will introduce TSP, VRP and several variants, present mathematical programming formulations and give some examples to the use of such formulations.

4.2 Travelling salesman problem

The TSP involves determining a minimum-cost route that starts from a depot, visits each customer exactly once from a given set of customers and then returns to the depot, with the additional restriction that each customer needs to be visited exactly once. The cost between a pair of nodes is often defined to be equal to the distance between the nodes, in which case the problem seeks a minimum-length route. There are no further restrictions such as capacities, or time windows, making the TSP one of the fundamental problems at the heart of distribution planning. Although simple to state, the problem is very difficult to solve to optimality. To substantiate this claim, consider a TSP with five customers to visit, which implies that there exist $5! = 120$ possible tours to choose from, in order to find the one with the minimum total length. The number of possible tours grows exponentially, where 10 customers result in just over 3.6 million possible tours and 20 customers as much as 2.43×10^{18}. The original problem dates back to the work of the mathematicians Hamilton and Kirkman, who formulated the mathematical problem for the first time in the 1800s. Sometimes, the TSP is also posed as finding a minimum cost Hamiltonian cycle on a given set of nodes.

The following example illustrates an application of the TSP.

Example 4.1

A courier operating in the south of the United Kingdom has received five 'next day delivery' small parcels, each to be delivered to a different location. The size, shape and weight of the parcels is such that it will suffice to dispatch one small van for the delivery operation. The van will leave at noon from the main depot based in Southampton, with deliveries to be made to Romsey, Winchester, Salisbury, Lyndhurst and Micheldever, to a location in the city or the town centre. The van will have to return back to the depot as soon as the deliveries are made, so that it can be ready for the next set of parcels. Traffic information indicates that there are no incidents, roadworks, congestion or blockages on the roads, and none expected for the rest of the day, which suggests that the time required to travel between two locations, for practical purposes, can be assumed to be proportional to the distance between the locations. The objective is to complete the deliveries as quickly as possible. A map of the geographical area in which the five locations are indicated is shown in Figure 4.1.

We will present two types of formulations in this section, starting with the asymmetric travelling salesman problem (ATSP) and then the symmetric travelling salesman problem (STSP).

4.2.1 Asymmetric formulations

To formulate the ATSP, we consider an asymmetric graph $G = (V, A)$ where $V = \{0, \ldots, n\}$ is the set of nodes including the depot indicated by node 0 and the set $V_C = \{1, \ldots, n\}$ of customers. We also use $A = \{(i, j) : i, j \in V, i \neq j\}$ to denote the set of arcs and c_{ij} is the cost (e.g. distance) associated

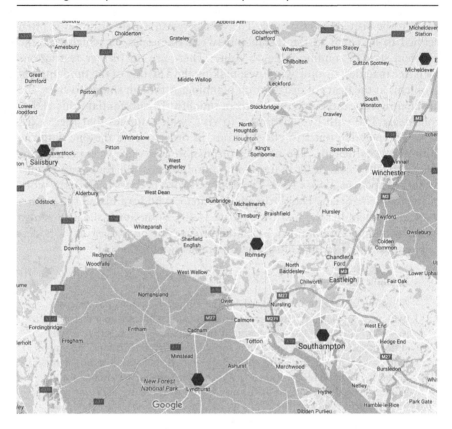

Figure 4.1 The geographical area of the instance in Example 4.1. (From Google Maps, 2016.)

with each arc $(i,j) \in A$. If the graph is incomplete, then there will be node pairs $i \in V$ and $j \in V\backslash\{i\}$, for which the arc (i,j) does not exist. In this case, the graph can be transformed into a complete network by introducing the arc (i,j) into the network, but with a sufficiently large value assigned to c_{ij}.

There are several ways to formulate the ATSP, but the most common one is to use an *assignment-based* formulation. The name stems from the fact that the formulation uses binary variables to assign each customer to two other nodes, one preceding and the other succeeding it. The variable, denoted by x_{ij}, is written for each arc $(i,j) \in A$ in the graph and is equal to 1 if arc (i,j) is used in an optimal tour and is equal to 0 otherwise.

$$\text{Minimise} \qquad \sum_{(i,j)\in A} c_{ij}x_{ij} \qquad\qquad (4.1)$$

subject to

$$\sum_{j\in V:(i,j)\in A} x_{ij} = 1 \qquad \forall i \in V \tag{4.2}$$

$$\sum_{i\in V:(i,j)\in A} x_{ij} = 1 \qquad \forall j \in V \tag{4.3}$$

Tour feasibility (4.4)

$$x_{ij} \in \{0, 1\} \quad \forall (i, j) \in A, \tag{4.5}$$

Constraints (4.2), (4.3) and (4.5) define the usual assignment relaxation for the ATSP. However, these two sets of constraints on their own are not always sufficient to model the problem, and additional constraints, referred to as *tour feasibility*, will be needed. The reasons for this will be explained in the following example, building on the application described in Example 4.1, following which we will present several ways to formulate tour feasibility.

Example 4.2

The TSP described in Example 4.1 is one of finding the shortest distance (or time, given the earlier assumption that they are proportional) route that starts from Southampton, visits all the five locations and returns to Southampton. The first piece of information needed to model the problem is the pairwise distances between the six locations specified, which is given in Table 4.1 where the following shorthand notation is used: Southampton (0), Romsey (1), Winchester (2), Salisbury (3), Lyndhurst (4) and Micheldever (5). The distances between all pairs of locations are extracted from an online map and are seen to exhibit an asymmetric structure. This is due to slight differences in road lengths, caused by one-way streets when entering and leaving the city or town centres.

Table 4.1 Distance data (in miles) for the parcel delivery example

	0	1	2	3	4	5
0	0	9.1	12.4	23.3	10.3	19
1	8.8	0	13.3	16.5	10.8	20.3
2	12.8	11.8	0	25.4	23.1	7.5
3	25.4	17.1	25	0	20.8	26.4
4	9.4	10.1	21.8	19.3	0	28.5
5	19.2	20.3	7.6	26.7	29.5	0

The next step in modelling the problem is to represent it using a graph with the nodes corresponding to the depot and the five delivery locations (or customers), and the links between the nodes. Since the distance matrix is asymmetric, the graph will be directed, meaning that for a pair of nodes there will be two arcs, in reverse directions. The resulting graph is depicted in Figure 4.2, which roughly corresponds to the geographical locations of the customers, but the distances are not to scale. Using the data and graph provided, a partial formulation for the problem using only the assignment constraints is given here. We will denote this formulation by $\mathcal{F}_a$:

Minimise

$$9.1x_{01} + 12.4x_{02} + 23.3x_{03} + 10.3x_{04} + 19x_{05}$$
$$+ 8.8x_{10} + 13.3x_{12} + 16.5x_{13} + 10.8x_{14} + 20.3x_{15}$$
$$+ 12.8x_{20} + 11.8x_{21} + 25.4x_{23} + 23.1x_{24} + 7.5x_{25}$$
$$+ 25.4x_{30} + 17.1x_{31} + 25x_{32} + 20.8x_{34} + 26.4x_{35}$$
$$+ 9.4x_{40} + 10.1x_{41} + 21.8x_{42} + 19.3x_{43} + 28.5x_{45}$$
$$+ 19.2x_{50} + 20.3x_{51} + 7.6x_{52} + 26.7x_{53} + 29.5x_{54}$$

subject to

$$x_{01} + x_{02} + x_{03} + x_{04} + x_{05} = 1$$
$$x_{10} + x_{12} + x_{13} + x_{14} + x_{15} = 1$$
$$x_{20} + x_{21} + x_{23} + x_{24} + x_{25} = 1$$
$$x_{30} + x_{31} + x_{32} + x_{34} + x_{35} = 1$$
$$x_{40} + x_{41} + x_{42} + x_{43} + x_{45} = 1$$
$$x_{50} + x_{51} + x_{52} + x_{53} + x_{54} = 1$$
$$x_{10} + x_{20} + x_{30} + x_{40} + x_{50} = 1$$
$$x_{01} + x_{21} + x_{31} + x_{41} + x_{51} = 1$$
$$x_{02} + x_{12} + x_{32} + x_{42} + x_{52} = 1$$
$$x_{03} + x_{13} + x_{23} + x_{43} + x_{53} = 1$$
$$x_{04} + x_{14} + x_{24} + x_{34} + x_{54} = 1$$
$$x_{05} + x_{15} + x_{25} + x_{35} + x_{45} = 1$$
$$x_{ij} \in \{0, 1\} \quad \forall i, j = 0, \ldots, 5, i \neq j.$$

An optimal solution of $\mathcal{F}_a$ shown earlier yields an optimal value of 68.4 miles. The corresponding solution is shown in Figure 4.3.

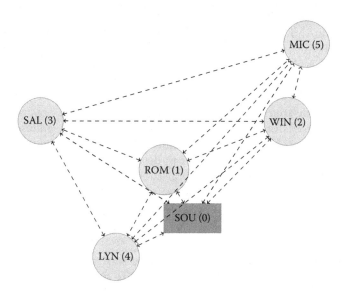

Figure 4.2 The fully connected graph describing the parcel delivery example (distances not to scale).

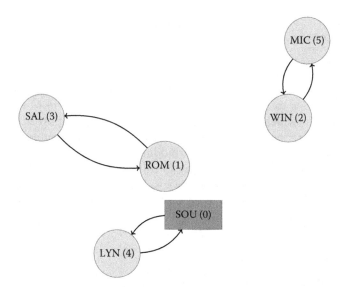

Figure 4.3 An optimal solution of formulation $\mathcal{F}_a$ with two subtours.

4.2.2 Subtours and subtour elimination for the ATSP

The solution shown in Figure 4.3 for the parcel delivery example satisfies the assignment constraints but does not constitute a feasible solution to the problem. This is because there are tours formed between customers that are disconnected from the depot. For example, the tour between nodes 1 and 3 starts from Romsey, visits Salisbury and returns back to Romsey. These infeasible tours are called *subtours*, and this is the reason why the assignment-based formulation (4.1) through (4.5) needs additional constraints shown by (4.4) so that any feasible solution of the formulation is free of subtours. There are different ways to describe such constraints for asymmetric graphs. We will present several such ways in this chapter, starting with the most intuitive ones, namely *cutset inequalities*.

4.2.2.1 Cutset inequalities

Before we present the cutset inequalities, we start by discussing a simpler version. Consider a subtour $\{i_1, i_2, i_3, \ldots, i_d, i_1\}$ that is formed between d customer nodes. This means that there are arcs in the solution $x^*_{i_1 i_2} = x^*_{i_2 i_3} = \cdots = x^*_{i_d i_1} = 1$ in an optimal solution x^* of the assignment-based formulation of the ATSP without any tour feasibility constraints imposed. These arcs induce a *circuit* C formed by the set of arcs $\{(i_1, i_2), (i_2, i_3), \ldots, (i_d, i_1)\}$. In this case, one can add the following so-called *circuit inequalities* to the formulation

$$\sum_{(i,j) \in C} x_{ij} \leq |C| - 1, \tag{4.6}$$

which ensure that this particular circuit C will be cut off from the set of feasible solutions to the problem. In particular, inequality (4.6) will forbid any solution in which all the arcs defined in the set C appear in the given sequence. The terms *cut-off*, *eliminate* or *break* are often interchangeably used when it comes to inequalities forbidding the formation of subtours. We will use these terms in the same way in this book. In the case of Example 4.1, the circuit inequalities for the two subtours shown in Figure 4.3 can be written as follows, one for each circuit defined on the node sets $\{1, 3\}$ and $\{2, 5\}$.

$$x_{13} + x_{31} \leq 1 \tag{4.7}$$

$$x_{25} + x_{52} \leq 1. \tag{4.8}$$

When formulation $\mathcal{F}_a$ is solved with these two circuit inequalities, one obtains the solution shown in Figure 4.4 with an optimal value of 82 miles,

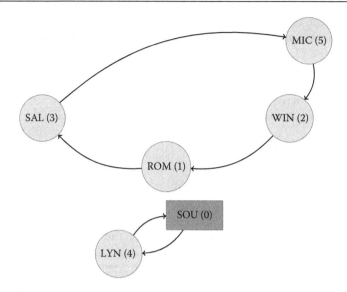

Figure 4.4 An optimal solution of formulation $\mathcal{F}_a$ with inequalities (4.7) and (4.8).

where it can be seen that the two subtours in the previous solution have now disappeared. However, another subtour, defined on a larger set $\{1,2,3,5\}$ of customers, has appeared. To break this subtour, it is possible to add the following circuit inequality to the formulation in a similar way to the previous iteration.

$$x_{13} + x_{35} + x_{52} + x_{21} \leq 3. \tag{4.9}$$

However, one can do better by adding *stronger* cuts to the formulation that will not only break a particular circuit $C = \{(i_1, i_2), (i_2, i_3), \ldots, (i_d, i_1)\}$ but will eliminate any subtour forming on the very set $\{i_1, i_2, \ldots, i_d\} \subseteq V_C$ of nodes defining the circuit. These inequalities are called *cutset* or *clique* inequalities proposed by Dantzig et al. (1954), written for all subsets V_S of the set V_C of customers, and are presented as follows:

$$\sum_{i \in V_S} \sum_{j \in V_S} x_{ij} \leq |V_S| - 1 \quad \forall V_S \subseteq V_C, |V_S| \geq 2. \tag{4.10}$$

Constraints (4.10) are exponential in number and ensure that the number of arcs which can be packed in the clique defined by the set of nodes V_S cannot exceed $|V_S| - 1$. For this reason, they are also referred to as *subtour elimination constraints* as they prohibit the formation of subtours on any subset V_S of the node set V_C. Interestingly, in the case of two-node subsets, as shown in

Figure 4.3, the circuit and clique inequalities are exactly the same. In other words, the circuit inequalities shown by (4.7) and (4.8) written for node sets {1, 3} and {2, 5}, respectively, are also the cutset (or clique) inequalities defined on the same set of nodes. However, this is not the case for subsets of three or more nodes. Coming back to our parcel delivery example, the cutset inequalities written for the subtour defined on the node set {1, 2, 3, 5} shown in Figure 4.4 would take the following form:

$$x_{13} + x_{31} + x_{35} + x_{53}$$
$$+ x_{25} + x_{52} + x_{12} + x_{21}$$
$$+ x_{15} + x_{51} + x_{23} + x_{32} \leq 3. \tag{4.11}$$

It is easy to see that inequalities (4.11) include the circuit inequalities (4.9) written for the same set of nodes and are therefore stronger. Indeed, an optimal solution of the formulation $\mathcal{F}_a$ with inequalities (4.7), (4.8) and (4.11) results in a solution shown in Figure 4.5, which is feasible in that all assignment constraints are satisfied, that no subtours are formed between the customers, and is therefore optimal with the objective equal to 83.7 miles.

An alternative way of writing the cutset inequalities is to consider that any subset V_S of the node set V_C will need to have at least one arc entering

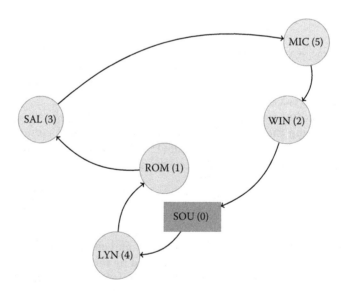

Figure 4.5 A feasible and an optimal solution of the parcel delivery example.

or leaving so that it will be connected to all the other nodes, which is shown in the following:

$$\sum_{i \in V_S} \sum_{j \in V_S'} x_{ij} \geq 1 \quad \forall V_S \subseteq V_C, |V_S| \geq 2, \tag{4.12}$$

where $V_S' = V \backslash V_S$ is a set to which V_S should be connected. For example, the subtour defined on nodes $\{1, 3\}$ shown in Figure 4.3, this alternative form of cutset inequalities would read as follows:

$$x_{10} + x_{12} + x_{14} + x_{15} + x_{30} + x_{32} + x_{34} + x_{35} \geq 1,$$

and similarly for the larger subtour shown in Figure 4.4 as follows:

$$x_{10} + x_{14} + x_{20} + x_{24} + x_{30} + x_{34} + x_{40} + x_{54} \geq 1.$$

Both representations of the cutset inequalities serve the same purpose of eliminating subtours, but given the structural differences in relation to the number variables with positive coefficients and the value of the right hand side, they might show a different performance when solved with off-the-shelf optimisers.

Based on the constraints introduced earlier, it is possible to derive a formulation of the ATSP as in the following, which we will refer to as the *cutset* formulation.

Minimise $\qquad \displaystyle\sum_{(i,j) \in A} c_{ij} x_{ij}$

subject to

$$\sum_{j \in V:(i,j) \in A} x_{ij} = 1 \qquad \forall i \in V$$

$$\sum_{i \in V:(i,j) \in A} x_{ij} = 1 \qquad \forall j \in V$$

$$\sum_{i \in V_S} \sum_{j \in V_S} x_{ij} \leq |V_S| - 1 \quad \forall V_S \subseteq V_C, |V_S| \geq 2$$

$$x_{ij} \in \{0, 1\} \qquad \forall (i, j) \in A.$$

Formulations such as *cutset* are sometimes referred to as *natural*, given that they are defined by the x_{ij} variables alone, a natural set of variables for this type of problems. In contrast, the so-called *extended* formulations use additional variables, the use of which we discuss and illustrate in the following sections.

4.2.2.2 Compact formulations

Solving the ATSP with the cutset inequalities (4.10) when the number of nodes is small and straightforward. In particular, for the five-customer parcel delivery problem introduced in Example 4.1, there will be a total of 26 of such constraints that can be appended to formulation $\mathcal{F}_a$. Some of these inequalities for subsets of size two and three are shown in Table 4.2. A solution to the formulation augmented with all such possible cutset inequalities will therefore always be feasible with respect to the assignment and tour constraints and yield an optimal solution to the problem.

While such an approach might be practical for problem instances with up to 10 or even 15 customers, the number of such constraints grows extremely quickly with larger instances. In this case, it would be more efficient to add them on an as-needed basis and in an iterative manner, in the same spirit of the approach taken in the previous section. This approach is called a cutting-plane algorithm, starting with solving an assignment-based formulation of the ATSP, adding subtour elimination constraints for any subtour that appears in the solution and therefore augmenting the formulation with such constraints at each iteration. This approach often requires implementation of special procedures to identify subtours that might appear in the intermediate solutions of the augmented formulation, the discussion of which is outside the scope of this book.

An alternative way to formulate the ATSP is to use so-called *compact* formulations that are polynomial in size. In other words, the number of

Table 4.2 Some cutset inequalities for $|V_S| = 2$ and $|V_S| = 3$ for the parcel delivery example

V_S	$|V_S|$	Cutset inequality (4.10)
$\{1, 2\}$	2	$x_{12} + x_{21} \leq 1$
$\{1, 3\}$	2	$x_{13} + x_{31} \leq 1$
$\{2, 3\}$	2	$x_{23} + x_{32} \leq 1$
$\{2, 4\}$	2	$x_{24} + x_{42} \leq 1$
$\{3, 4\}$	2	$x_{34} + x_{43} \leq 1$
$\{4, 5\}$	2	$x_{45} + x_{54} \leq 1$
$\{1, 2, 3\}$	3	$x_{12} + x_{21} + x_{13} + x_{31} + x_{23} + x_{32} \leq 2$
$\{1, 3, 4\}$	3	$x_{13} + x_{31} + x_{14} + x_{41} + x_{34} + x_{43} \leq 2$
$\{1, 3, 5\}$	3	$x_{13} + x_{31} + x_{15} + x_{51} + x_{35} + x_{53} \leq 2$
$\{2, 3, 4\}$	3	$x_{23} + x_{32} + x_{34} + x_{43} + x_{24} + x_{42} \leq 2$
$\{2, 4, 5\}$	3	$x_{24} + x_{42} + x_{25} + x_{52} + x_{45} + x_{54} \leq 2$
$\{3, 4, 5\}$	3	$x_{34} + x_{43} + x_{35} + x_{53} + x_{45} + x_{54} \leq 2$

constraints required in such formulations is a polynomial function of the number of nodes or arcs in the instance. Compact formulations for the ATSP require the use of auxiliary variables and therefore require an increase in the number of variables in the formulation over natural formulations. However, the use of such variables allows capturing additional information about the problem which otherwise would not be possible or straightforward, as discussed in the following.

4.2.2.3 Compact node-ordering formulation

One type of compact formulation for the ATSP uses additional variables $u_i \geq 0$ for each node $i \in V_C$ that specifies the position of customer i on the tour, for an instance of the ATSP with a total of $n + 1$ nodes, including the depot (i.e. $|V_C| = n$). The new variables are such that if an arc $(i, j) \in A$ is used in the solution, then the values of new variables should satisfy the condition $u_j = u_i + 1$, meaning that the position of customer j should be the one immediately after node i. This information can be conveyed using the following constraints:

$$u_i - u_j + nx_{ij} + (n - 2)x_{ji} \leq n - 1 \quad \forall i, j \in V_C, i \neq j \tag{4.13}$$

It is easy to see that when $x_{ij} = 1$ for a particular arc (i, j), then constraints (4.13) read as follows,

$$u_i - u_j + n \leq n - 1 \quad \text{written for arc } (i, j) \tag{4.14}$$

$$u_j - u_i + n - 2 \leq n - 1 \quad \text{written for arc } (j, i), \tag{4.15}$$

which jointly imply $u_j = u_i + 1$. Alternatively, if a particular arc $(i, j) \in A$ or its reverse $(j, i) \in A$ has not been used in the solution, then this implies $x_{ij} = x_{ji} = 0$ and the constraints read:

$$u_i - u_j \leq n - 1 \quad \text{written for arc } (i, j) \tag{4.16}$$

$$u_j - u_i \leq n - 1 \quad \text{written for arc } (j, i), \tag{4.17}$$

which effectively allows variables u_i and u_j to take any values such that the difference is at most $n - 1$. An implication of this condition is that the position of the first node on the tour $i \in V_C$ can be as small as $u_i = 0$, and the position of the last node j on the tour would then be $u_j = n - 1$, leading to a difference that is bounded by $n - 1$. When constraints (4.13) replace constraints (4.4) in the ATSP model (4.1) through (4.5), a valid formulation is obtained, which we will henceforth refer to as a *compact node-ordering formulation*. Constraints (4.13) originate from the work of Miller et al. (1960)

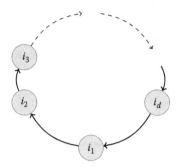

Figure 4.6 A subtour forming among customer nodes V_C.

and are proposed by Desrochers and Laporte (1991) as strengthenings of the former.

To see how inequalities (4.13) do not permit any subtour, consider a tour formed by the arcs $\{(i_1, i_2), (i_2, i_3), \ldots, (i_d, i_1)\}$ where $i_1, i_2, i_3, \ldots, i_d \in V_C$, as shown in Figure 4.6, which implies a solution $x_{i_1 i_2} = x_{i_2 i_3} = \cdots = x_{i_d i_1} = 1$. This solution, through the implications of constraints (4.14) and (4.15), yields the following set of equalities:

$$u_{i_2} = u_{i_1} + 1$$
$$u_{i_3} = u_{i_2} + 1 = u_{i_1} + 2$$
$$\ldots$$
$$u_{i_1} = u_{i_d} + 1 = \cdots = u_{i_1} + d - 1,$$

where the last equation is a contradiction. Therefore, a subtour such as the one shown here will never satisfy inequalities (4.13) and will therefore never form in a feasible solution of a compact node-ordering formulation.

An example illustrating the use of these constraints is given here.

Example 4.3

We consider the parcel delivery introduced in Example 4.1 and show how inequalities (4.13) can be used to model the resulting ATSP building on the assignment-based formulation. To this end, we define five additional continuous variables u_1, u_2, u_3, u_4 and u_5, which correspond to the position of nodes 1–5 in an optimal tour, respectively. The full set of constraints is as follows:

$$u_1 - u_2 + 5x_{12} + 3x_{21} \le 4$$
$$u_2 - u_1 + 5x_{21} + 3x_{12} \le 4$$

$$u_1 - u_3 + 5x_{13} + 3x_{31} \leq 4$$
$$u_3 - u_1 + 5x_{31} + 3x_{13} \leq 4$$
$$u_1 - u_4 + 5x_{14} + 3x_{41} \leq 4$$
$$u_4 - u_1 + 5x_{41} + 3x_{14} \leq 4$$
$$u_1 - u_5 + 5x_{15} + 3x_{51} \leq 4$$
$$u_5 - u_1 + 5x_{51} + 3x_{15} \leq 4$$
$$u_2 - u_3 + 5x_{23} + 3x_{32} \leq 4$$
$$u_3 - u_2 + 5x_{32} + 3x_{23} \leq 4$$
$$u_2 - u_4 + 5x_{24} + 3x_{42} \leq 4$$
$$u_4 - u_2 + 5x_{42} + 3x_{24} \leq 4$$
$$u_2 - u_5 + 5x_{25} + 3x_{52} \leq 4$$
$$u_5 - u_2 + 5x_{52} + 3x_{25} \leq 4$$
$$u_3 - u_4 + 5x_{34} + 3x_{43} \leq 4$$
$$u_4 - u_3 + 5x_{43} + 3x_{34} \leq 4$$
$$u_3 - u_5 + 5x_{35} + 3x_{53} \leq 4$$
$$u_5 - u_3 + 5x_{53} + 3x_{35} \leq 4$$
$$u_4 - u_5 + 5x_{45} + 3x_{54} \leq 4$$
$$u_5 - u_4 + 5x_{54} + 3x_{45} \leq 4$$

When this set of constraints are appended to formulation $\mathcal{F}_a$ shown in Example 4.1, one obtains an optimal solution shown in Figure 4.7, where the optimal values u_i^* of the auxiliary variables are shown next to each node $i = 1, \ldots, 5$, indicating the visit order of the customers.

4.2.2.4 Compact arc-ordering or flow-based formulation

The node-ordering formulation presented in the previous section assigns position numbers to *nodes*, which correspond to the order in which they are visited. In a similar way, an alternative compact formulation can be described that prescribes the order in which the *arcs* are visited in the tour. In particular, let f_{ij} be a continuous variable representing the position of arc $(i, j) \in A$ if this arc is traversed in a tour. Similar to the interpretation of the u_i variables in the previous section, the f_{ij} variables should be such that $f_{jk} = f_{ij} + 1$ if $x_{ij} = x_{jk} = 1$. In other words, for three distinct nodes $i, j, k \in V$, if two arcs are adjacent to each other, then the difference in flow from one to another should be equal to 1. This situation is illustrated in Figure 4.8,

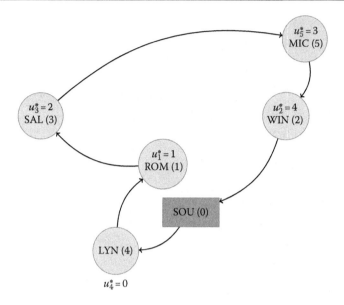

Figure 4.7 An optimal solution of the parcel delivery example when solved using inequalities (4.13) to prevent subtours.

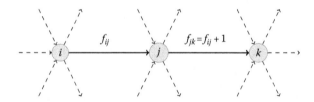

Figure 4.8 Illustration of the way that the flow variables model the arc positions.

where the dashed arcs are those that are part of the graph but are not used in the solution and for which the corresponding f_{ij} variables have a value equal to zero. The solid arcs in the graph are the ones that are part of the given tour and which would be assigned a nonnegative value depending on the order of their visit.

The following set of constraints can be used to describe the relationship between the f_{ij} and x_{ij} variables when modelling the ATSP defined on $n+1$ nodes including the depot:

$$\sum_{j\in V\setminus\{i\}} f_{ij} - \sum_{j\in V\setminus\{i\}} f_{ji} = 1 \qquad \forall i \in V_C \tag{4.18}$$

$$f_{ij} \le nx_{ij} \quad \forall (i,j) \in A. \tag{4.19}$$

Constraints (4.18) model the difference between the positions of successive arcs in a tour, and constraints (4.19) simply state that an arc $(i, j) \in A$ that does not appear in a tour (i.e. $x_{ij} = 0$) cannot be assigned a position.

Sometimes an analogy is made between the f_{ij} variables and the distribution of a single commodity on graph G, in that the variables are interpreted as modelling the *flow* of the commodity. With this interpretation, constraints (4.18) would correspond to the case where, upon visiting each customer node, one unit of commodity is delivered. The total amount of commodity distributed would then be n, equal to the number of customers, for which reason the same value is used as an upper bound on the auxiliary variables in constraint (4.19). For this reason, the f_{ij} variables are sometimes referred to as *flow variables*. Formulation (4.1) through (4.5), where (4.4) are replaced by constraints (4.18) and (4.19), is known as the *single-commodity flow* formulation of the ATSP, described as early as 1978 in the work by Gavish and Graves (1978). An example illustrating the use of these constraints is given next.

Example 4.4

A single-commodity flow formulation of the parcel delivery instances described in Example 4.2 would use an additional 30 continuous flow variables, one corresponding to each arc $(i, j) \in A$ of the network, for which some of the constraints that would appear in the single-commodity flow formulation are shown here:

$$f_{10} + f_{12} + f_{13} + f_{14} + f_{15} - f_{01} - f_{21} - f_{31} - f_{41} - f_{51} = 1$$
$$f_{20} + f_{21} + f_{23} + f_{24} + f_{25} - f_{02} - f_{12} - f_{32} - f_{42} - f_{52} = 1$$
$$f_{30} + f_{31} + f_{32} + f_{34} + f_{35} - f_{03} - f_{13} - f_{23} - f_{43} - f_{53} = 1$$
$$f_{40} + f_{41} + f_{42} + f_{43} + f_{45} - f_{04} - f_{14} - f_{24} - f_{34} - f_{54} = 1$$
$$f_{50} + f_{51} + f_{52} + f_{53} + f_{54} - f_{05} - f_{15} - f_{25} - f_{35} - f_{45} = 1$$
$$f_{10} \leq 5x_{10}$$
$$f_{20} \leq 5x_{20}$$
$$f_{30} \leq 5x_{30}$$
$$\ldots$$
$$f_{25} \leq 5x_{25}$$
$$f_{35} \leq 5x_{35}$$
$$f_{45} \leq 5x_{45}$$

An optimal solution obtained by the single-commodity flow formulation using these sets of constraints is shown in Figure 4.9, where the optimal values f_{ij}^* of the auxiliary variables are shown next to each arc in the graph.

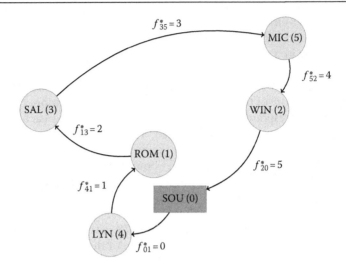

Figure 4.9 An optimal solution of the parcel delivery example when solved using the single-commodity flow formulation.

4.2.2.5 Compact precedence-based formulation

The compact formulations introduced so far use one auxiliary variable that is defined individually for each element (either node or arc) of the network and relate these elements, using additional sets of constraints. For example, the compact node-ordering formulation uses one auxiliary variable u_i for each node $i \in V_C$, and the additional constraints ensure that variables u_i and u_j are related whenever arcs (i, j) or (j, i) appear in a solution. The same concept applies to the compact arc-ordering formulation, with the only difference that the variables are individually defined for the arcs.

A third way to model subtour elimination in a compact manner is to use, once again, a set of auxiliary variables, but where the variables involve pairs of nodes. The new variables in this case would model the precedence relationship between the two nodes. It should be remembered that the two-index natural variables x_{ij} already define a precedence relationship between two nodes $i \in V_C$ and $j \in V_C\backslash\{j\}$, but the precedence used here is immediate in the sense that if $x_{ij} = 1$, then node i is the node visited just before node j. In contrast, the new auxiliary variables in the compact formulation we describe here model the precedence relationship in a slightly different manner. In particular, a binary variable v_{pi} is equal to 1 if node $p \in V_C$ is in the path between the depot (node 0) and node $i \in V_C\backslash\{p\}$ and is equal to 0 otherwise. Figure 4.10 shows a path of a given route that leaves the depot and visits four customer nodes $\{i, j, k, l\}$, on which the new auxiliary variables that take the value 1 for this particular path are indicated.

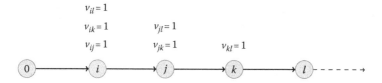

Figure 4.10 Illustration of the precedence-based auxiliary variables.

The following set of constraints shows the relationship between the natural variables and the new set of auxiliary variables and breaks any subtours forming among the customer nodes:

$$x_{ij} \leq v_{ij} \qquad \forall i,j \in V_C, i \neq j \tag{4.20}$$

$$x_{ij} + v_{ji} \leq 1 \qquad \forall i,j \in V_C, i \neq j \tag{4.21}$$

$$x_{ij} + v_{pi} \leq v_{pj} + 1 \quad \forall i,j,p \in V_C, i \neq j \neq p. \tag{4.22}$$

It can easily be seen that the new variables model the precedence between nodes p and i in a more general manner, as compared to the natural variables, that is, if $x_{pi} = 1$ then this implies that $v_{pi} = 1$ must hold, but not the other way around. Constraints (4.20) model precisely this condition. The next set of constraints (4.21) indicates that if arc (i,j) appears in an optimal solution (i.e. $x_{pi} = 1$), indicating that node $j \in V_C$ must immediately follow node $i \in V_C\backslash\{j\}$ in the tour, then it must be that $v_{ji} = 0$, namely node j cannot appear on a path between the depot and node i. The last set (4.22) of constraints is written for all triplets of customer nodes and ensures that the precedence information is consistent across all the nodes. More specifically, these constraints stipulate that if a node $p \in V_C$ is in the path between the depot and a node $i \in V_C\backslash\{p\}$, and if the arc (i,j) is traversed in a solution, then the node p must also be in the path between the depot and node $j \in V_C$, $j \neq i, j \neq p$.

To see why constraints (4.20) through (4.22) prohibit the formation of subtours, consider Figure 4.11 which shows an infeasible tour formed among three customers $\{i,j,k\}$ not involving the depot.

According to the subtour in Figure 4.11, $x_{ij} = x_{jk} = x_{ki} = 1$, which implies, through constraints (4.20), $v_{ij} = v_{jk} = v_{ki} = 1$. In this case, constraints (4.21) imply $v_{ji} = v_{kj} = v_{ik} = 0$. However, these values violate the inequalities (4.22) written for the node triplet i,j,k in the form $x_{ij} + v_{ki} \leq v_{kj} + 1$, yielding $2 \leq 1$, which is clearly infeasible. This result can be generalised to show that subtours containing more than three nodes also result in the same infeasibility.

One feature of the set of inequalities (4.20) through (4.22) is that they are valid even if the integrality restriction on the auxiliary variables is relaxed

Figure 4.11 A three-node subtour.

as $v_{ip} \geq 0$, which implies that the use of such constraints does not result in an increase in the number of integer variables in the formulation. Another interesting aspect is that when aggregated, constraints (4.20) through (4.22) imply a weaker form of the constraints (4.13) and are therefore disaggregations of the latter. In fact, it can easily be seen that the following relationship between the auxiliary variable u_i of the node-ordering formulation and those of the precedence-base formulation holds,

$$\sum_{p \in V_C \setminus \{i\}} v_{pi} = u_i \quad \forall i \in V_C,$$

which again shows that the latter variables are disaggregations of the former ones.

We conclude this section by noting that constraints (4.22) can be strengthened as follows:

$$x_{ij} + x_{ji} + v_{pi} \leq v_{pj} + 1 \quad \forall i, j, p \in V_C, i \neq j \neq p$$

which are shown to be disaggregations of constraints (4.13). Finally, an alternative way of strengthening constraints (4.22) is as follows:

$$x_{ij} + x_{pj} + x_{ip} + v_{pi} \leq v_{pj} + 1 \quad \forall i, j, p \in V_C, i \neq j \neq p.$$

For further details on these constraints, the reader is referred to Gouveia and Pires (1999, 2001).

Example 4.5

In the following, we show how some of the constraints (4.20) through (4.22) would appear when applied to Example 4.1. Consider, for example, the arc $(1, 2)$ in this instance, for which the three sets of inequalities (4.20) through (4.22) would appear as follows:

$$x_{12} \leq v_{12}$$

$$x_{12} + v_{21} \leq 1$$

$$x_{12} + v_{31} \leq v_{32} + 1 \quad p = 3$$

$$x_{12} + v_{41} \leq v_{42} + 1 \quad p = 4$$

$$x_{12} + v_{51} \leq v_{52} + 1 \quad p = 5.$$

Similarly, for arc $(2, 1)$, the constraints would read as follows:

$$x_{21} \leq v_{21}$$

$$x_{21} + v_{12} \leq 1$$

$$x_{21} + v_{32} \leq v_{31} + 1 \quad p = 3$$

$$x_{21} + v_{42} \leq v_{41} + 1 \quad p = 4$$

$$x_{21} + v_{52} \leq v_{51} + 1 \quad p = 5.$$

For this six-node instance, there would be $5 \times 5 - 5 = 20$ of constraints (4.20), written for all pairs of customer nodes. A similar number applies to the number of constraints (4.21). Finally, there would be $20 \times 3 = 60$ of constraints (4.22).

4.2.3 Symmetric formulations

If the underlying network of the TSP indicates a symmetric structure, the resulting problem is the STSP, a special case of the ATSP. This observation indicates that the formulations presented in the previous section are also applicable for the STSP, which would still assume a graph where there are two arcs, in opposite directions, between a pair of nodes, that replace the edge. However, one can derive more efficient formulations for the STSP in relation to the number of binary variables needed, given that an edge between a pair of nodes requires a single variable to indicate whether it has been traversed or not. We will elaborate more on this in the following.

In more formal terms, the STSP can be modelled on a graph $G = (V, E)$ in which $V = \{0, 1, \ldots, n\}$ is the set of nodes as defined in the previous section and $E = \{\{i, j\} | i, j \in V, i < j\}$ is the set of edges. Each edge $\{i, j\} \in E$ has a weight $c_{\{ij\}}$, for example distance between nodes i and j. In this case, we define the decision variable $x_{\{ij\}}$ which takes the value 1 if edge $\{i, j\}$ is used, and 0 otherwise. A natural formulation using cutset inequalities for the STSP is as follows:

$$\text{Minimise} \qquad \sum_{\{i,j\}\in E} c_{\{ij\}} x_{\{ij\}} \tag{4.23}$$

subject to

$$\sum_{j\in V:\{i,j\}\in E} x_{ij} = 2 \qquad \forall i \in V \tag{4.24}$$

$$\sum_{i\in V_S} \sum_{j\in V\backslash V_S} x_{\{ij\}} \geq 2 \qquad \forall V_S \subseteq V_C, 2 \leq |V_S| \leq n/2 \tag{4.25}$$

$$x_{\{ij\}} \in \{0,1\} \quad \forall \{i,j\} \in E. \tag{4.26}$$

In this formulation, the objective function (4.23) minimises the total routing cost. Constraints (4.24) guarantee that each node is adjacent to exactly two edges. Possible subtours forming among nodes in the set V_C are eliminated by constraints (4.25) which are written for $|V_S| \geq 2$ and $|V_S| \leq n/2$. The reason for the former condition is that the case $|V_S| = 1$ is already represented by constraints (4.24) for single nodes. As for the latter, it suffices to write these constraints for only half the number of possible subsets, given the symmetry. In particular, it is easy to see that

$$\sum_{i\in V_S} \sum_{j\in V\backslash V_S} x_{\{ij\}} = \sum_{i\in V\backslash V_S} \sum_{j\in V_S} x_{\{ij\}},$$

which shows that constraints (4.25) model the same situation when written separately for subsets V_S and $V\backslash S$. The last set of constraints (4.26) model the binary requirements on the decision variables. For a STSP defined on $n + 1$ nodes including the depot, the earlier formulation uses $(n + 1)^2/2 - (n + 1) = n^2 - 1$ binary variables for a fully connected graph, as opposed to what would have been needed in a formulation of the ATSP, which is $n^2 + n$. Similar to the ATSP, constraints (4.25) are exponential in number, as sets V_S would need to be defined for each subset of the set V_C of customer nodes.

The use of this formulation is illustrated in Example 4.6, this time for an incomplete graph.

Example 4.6

Consider the small-scale STSP instance shown in Figure 4.12 defined on a graph with five nodes, where the depot is shown by node 0 and the customers are shown by nodes 1–4.

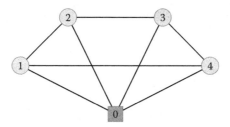

Figure 4.12 A symmetric and incomplete graph defined on five nodes.

For this instance, constraints (4.24), written separately for each node 0–4, would read as follows:

Node 0: $x_{\{01\}} + x_{\{02\}} + x_{\{03\}} + x_{\{04\}} = 2$

Node 1: $x_{\{01\}} + x_{\{12\}} + x_{\{14\}} = 2$

Node 2: $x_{\{02\}} + x_{\{12\}} + x_{\{23\}} = 2$

Node 3: $x_{\{03\}} + x_{\{23\}} + x_{\{34\}} = 2$

Node 4: $x_{\{04\}} + x_{\{14\}} + x_{\{34\}} = 2.$

As for the subtour elimination constraints (4.25), we provide some examples here:

For $V_S = \{1, 2\}$:

$x_{\{01\}} + x_{\{02\}} + x_{\{23\}} + x_{\{14\}} \geq 2$

For $V_S = \{1, 3\}$:

$x_{\{01\}} + x_{\{03\}} + x_{\{12\}} + x_{\{14\}} + x_{\{23\}} + x_{\{34\}} \geq 2$

For $V_S = \{1, 2, 3\}$:

$x_{\{01\}} + x_{\{02\}} + x_{\{03\}} + x_{\{14\}} + x_{\{34\}} \geq 2$

Consider, now, the following set of subtour elimination constraints written for $V_S = \{0, 2, 4\}$, where $V_S \subset V$, as opposed to $V_S \subseteq V_C$:

$x_{\{01\}} + x_{\{03\}} + x_{\{12\}} + x_{\{14\}} + x_{\{23\}} + x_{\{34\}} \geq 2$

As can be seen from these inequalities, the subtour elimination constraints written for subsets $\{1, 3\}$ and $\{0, 2, 4\}$ are exactly the same, given the symmetry. In this case, it suffices to include only the former in the formulation for this particular subset configuration.

4.3 Vehicle routing problem

The TSP discussed in the previous section relates to finding a single route, as might be the case in a single van delivering parcels to a number of customers or a ship visiting several ports. However, most real-life freight distribution problems, especially those arising on large networks involving tens or hundreds of delivery points, require the routing of a fleet of vehicles. This consideration alone extends the TSP into what is widely known as the VRP, a general name for a class of problems incorporating realistic and complex constraints. In its basic form, the VRP consists of finding least-cost vehicle routes for serving a number of customers, each requiring pick up or delivery of a known amount of goods, such that each customer is visited exactly once by a vehicle and that all vehicle routes start from and end at a depot. Multiple visits to a given customer by a given vehicle (as in the case of split deliveries) or visits to a customer by different vehicles are not allowed in this basic version, although there are variants of the VRP which relax these restrictions, as we will see later in the chapter. The basic version also assumes a single depot and a homogeneous fleet of vehicles, where these terms were discussed earlier on in the chapter.

The VRP was introduced over 50 years ago by Dantzig and Ramser (1959), who studied the problem concerning the 'optimum routing of a fleet of gasoline delivery trucks between a bulk terminal and a large number of service stations supplied by the terminal', which they initially named as the 'truck dispatching problem'. Using a method based on linear programming, their hand calculations produced a near-optimal solution with four routes to a problem involving 12 service stations. A large number of practical applications of the problem have since been described, in which much larger instances of the problem have been successfully solved.

The VRPs that arise in real-life applications rarely appear in their basic form. The three commonly studied variants are as follows:

1. Each vehicle has a limited capacity and the total demand served on a vehicle route cannot exceed this capacity. This version of the VRP is named the *capacitated vehicle routing problem* (CVRP).
2. Each vehicle is limited to a maximum length, which cannot be exceeded in a given tour. This version of the VRP is known as the *distance-constrained vehicle routing problem* (DVRP).
3. Each customer requests to be served in a given time interval defined by the earliest time and latest time, which is referred to as the *vehicle routing problem with time windows* (VRPTW).

There also exist combinations of these problems, such as the capacitated vehicle routing problem with time windows (CVRPTW) and the distance- and capacity-constrained vehicle routing problem (DCVRP), among others.

The basic VRP can be modelled in several ways, and the choice of model to use depends on the level of detail to be captured as well as the complexity of the resulting formulation. We will present three types of formulations in the following sections, which differ with respect to the type of the natural variable used to model the problem. The first of these is the *two-index formulation*, similar to the assignment-based formulation of the TSP, where a decision variable is defined for each arc traversed in the network, irrespective of the particular vehicle used on each arc. The second assignment-based formulation explicitly models the route of each vehicle in the fleet, which will require an additional index in the decision variable, and for this reason will be referred to as the *three-index formulation*. Finally, we will look at a different type of a model that breaks away from the assignment-based paradigm and instead uses set partitioning approaches to formulate the problem. Before presenting these models in detail, we will first introduce the notation that is common to all formulations. The rest of the exposition will relate to asymmetric graphs, as it also covers symmetric graphs. We will show later how these formulations can be extended to accommodate the variants of the VRP shown earlier, such as the CVRP and the VRPTW.

4.3.1 Common notation

The basic VRP is modelled on an asymmetric graph $G = (V, A)$ where $V = \{0, \ldots, n\}$ is the set of nodes including the depot indicated by node 0, and $V_C = \{1, \ldots, n\}$ is the set of customers. As in the TSP, $A = \{(i, j) : i, j \in V, i \neq j\}$ denotes the set of arcs and c_{ij} denotes the cost (e.g. distance) associated with each arc $(i, j) \in A$. There exists a set K of vehicles with the following attributes:

- The fleet is either limited in number in which case we denote the total number of vehicles by m, or sufficiently many where it is possible to use as many vehicles as required, but the use of each vehicle $k \in K$ will incur a fixed cost g^k.
- Each vehicle has limited capacity denoted by π^k, where capacity will be treated as being uni-dimensional (e.g. in kilograms or tonnes), in relation to carrying a single type of a product. As the basic VRP assumes a homogeneous fleet, $\pi^k = \pi$ for all $k \in K$.
- A maximum tour length τ^k is imposed on each vehicle $k \in K$, and where this is a common limitation across the fleet, this will be shown as $\tau^k = \tau$ for all $k \in K$.
- It is implicitly assumed that each vehicle can be used at most once during the planning period; in other words, the multiple use of vehicles in a given period (e.g. day) is not allowed. However, there are variants of the basic VRP where this assumption is relaxed, where a vehicle can return to the depot as many times as feasible to replenish its inventory and continue to serve the customers.

The customers have the following attributes:

- Each customer $i \in V_C$ has a positive demand for the product that is shown by q_i. Any customer $i \in V_C$ with demand $q_i = 0$ can be removed from the problem. Similarly, any customer $i \in V_C$ whose demand exceeds the vehicle capacity, that is $q_i > \pi$, one can pre-allocate $\lceil (q_i - q_i \bmod \pi)/\pi \rceil$ dedicated vehicles to this customer *a priori* and only consider any positive 'excess' demand $q_i = q_i \bmod \pi$ in the problem. In other words, the amount of demand in this case is first met with as many full truckload shipments as possible, and any leftover amount that is smaller than the common vehicle capacity is included in the problem. Although demands are only attributed to customers, sometimes it is convenient to define $q_0 = 0$ for modelling purposes.
- Each customer $i \in V_C$ is associated with a service time window $[a_i, b_i]$, where a_i is the earliest possible time that service can commence and b_i is the latest time at which service must commence. The length of time it takes to serve this customer, or the service time, is denoted by s_i. If deviations from the time window are allowed, then this will be at the expense of penalties.

Example 4.7

A supplier of fresh produce has arranged deliveries to be made to the five locations shown in Figure 4.2 from the depot located in Southampton on a given day of the week. The supplier has access to a fleet of five identical vans, each of which has space to accommodate a maximum of four pallets. The supplier does not necessarily need to use all five vans. The amounts requested by the five locations are two pallets each for Salisbury and Romsey, three pallets each for Micheldever and Lyndhurst and one pallet for Winchester. The depots where the deliveries are to be received will be open from 9am to 6pm on the day of delivery. Each delivery will take around 15 minutes to be unloaded from the van and taken into the depot, including the signing of the paperwork. This is an example of a CVRP where the demands are represented by $(q_1, q_2, q_3, q_4, q_5) = (2, 1, 2, 3, 3)$ and the common vehicle capacity is $\pi = 4$. It can be seen that although the delivery locations have a common time window represented by [9am, 6pm], these will not pose a restriction as will be shown later. Furthermore, there are no constraints relevant to maximum tour length or maximum time of continuous driving for the drivers.

This example can also be used to illustrate the situation for when the demand for one of the locations exceeds the vehicle capacity. If the demand of, say, Lyndhurst was specified as nine pallets, then $\lceil (9 - 9 \bmod 4)/4 \rceil = 2$ additional vans from the fleet can be sent to this location directly to pick up eight of the pallets and leave the remaining $9 - 4 \times 2 = 1$ pallet to be picked up by one of the remaining vehicles, possibly consolidated with other orders. In contrast, if the demand was

$q_4 = 8$, then the demand in Lyndhurst will have to be served with two full truckloads and it will not be possible to combine any shipment for this node with shipment for any other node. In this case, Lyndhurst can be removed from the problem input data.

4.4 Capacitated vehicle routing

CVRP is the most basic variant of the family of vehicle routing problems and aims to find a minimum-cost set of routes for a homogeneous fleet of vehicles to serve a set of customers, such that each route starts and ends at the depot and that the vehicles carry at most π units of commodity. For this problem, any arc $(i, j) \in A$ such that $q_i + q_j > \pi$ is infeasible and will never appear in a feasible solution to the problem. All such arcs can be removed from graph G, prior to formulating the problem.

A variety of formulations have been described for the CVRP. Three of these will be presented in the following.

4.4.1 Two-index formulations

The two-index formulation extends the assignment-based model described for the ATSP and takes its name from the fact that one binary variable x_{ij} with two indices is associated with each arc $(i, j) \in A$. The binary variable is equal to 1 if arc (i, j) appears in a tour, and 0 otherwise. The general structure of the two-index formulation for the VRP using all $|K|$ vehicles is presented as follows:

$$\text{Minimise} \quad \sum_{(i,j) \in A} c_{ij} x_{ij} \tag{4.27}$$

subject to

$$\sum_{j \in V_C : (i,j) \in A} x_{0j} = |K| \tag{4.28}$$

$$\sum_{i \in V_C : (i,j) \in A} x_{i0} = |K| \tag{4.29}$$

$$\sum_{j \in V : (i,j) \in A} x_{ij} = 1 \qquad \forall i \in V_C \tag{4.30}$$

$$\sum_{i \in V : (i,j) \in A} x_{ij} = 1 \qquad \forall j \in V_C \tag{4.31}$$

Tour feasibility $\tag{4.32}$

$$x_{ij} \in \{0, 1\} \quad \forall (i, j) \in A. \tag{4.33}$$

In this formulation, constraints (4.28) and (4.29) ensure that there are exactly $|K|$ vehicles that start from and return to the depot. Constraints (4.30) and (4.31) impose the assignment restrictions on the customer nodes.

If the number of vehicles is not fixed or bounded by a constant $\bar{m}$, then they can be replaced by the following set of constraints:

$$\sum_{j \in V_C : (i,j) \in A} x_{0j} \leq \bar{m}$$

$$\sum_{i \in V_C : (i,j) \in A} x_{i0} \leq \bar{m}.$$

Finally, if there are no limits on the number of vehicles that can be used, but the use of a vehicle incurs a common fixed cost g, then constraints (4.28) and (4.29) can be removed from the formulation. In this case, the objective function would be augmented as follows:

$$\text{Minimise} \qquad \sum_{(i,j) \in A} c_{ij} x_{ij} + g \sum_{j \in V_C} x_{0j}.$$

Constraints (4.32) describe the tour feasibility requirements. In contrast to the TSP, tour feasibility in the VRP might pertain to a number of conditions, and not just connectivity. For example, tour feasibility for the CVRP implies that each tour should start from and end at the depot (in other words, connectivity) and satisfy the capacity limitations. For the DVRP, tour feasibility includes connectivity as well as the length of each tour staying within the given threshold. In the following, we will present three different ways in which constraints (4.32) can be written.

4.4.1.1 Cutset inequalities for tour feasibility

Similar to the TSP, cutset inequalities can be written for the VRP in the natural space of variables. In the case of the CVRP, only one set of such constraints is sufficient to ensure that both connectivity and the capacity limitations on each route are respected. These constraints are based on the simple idea that the number of vehicles serving any subset V_S of customer nodes should be sufficient to meet the demand $q(V_S) = \sum_{i \in V_S} q_i$ of that subset. Let $b(V_S)$ denote the minimum number of vehicles required to serve this demand. Then, the cutset inequalities are presented as follows:

$$\sum_{i \in V_S, j \in V'_S} x_{ij} \geq b(V_S), \qquad \forall V_S \subseteq V_C, V_S \neq \emptyset \qquad (4.34)$$

where $V'_S = V \backslash V_S$. Determining $b(V_S)$ can be done by solving another combinatorial optimisation problem, namely *bin packing* (see, e.g., Martello and Toth, 1990), which is a difficult task in itself. In practice, a quick way is to approximate this value by $\lceil q(V_S)/\pi \rceil$, which does not necessarily equal $b(V_S)$ but acts as a lower bound instead, as the following example will show.

Example 4.8

Consider a CVRP with three customers, each of which has a demand equal to two units, and there are a sufficient number of identical vehicles available with a capacity $\pi = 3$ each. The value $\lceil 6/3 \rceil = 2$ implies that *at least* two vehicles are needed to serve this demand. In contrast, the solution to the bin packing problem indicates that at least three vehicles are needed, given that it is not possible to combine the demands of two customers in a single vehicle.

It is easy to see how constraints (4.34) work by observing that an infeasible subset V_S with $q(V_S) > \pi$ implies $b(V_S) \geq \lceil q(V_S)/\pi \rceil > 1$, indicating that at least two arcs should be leaving subset V_S, in other words, at least two routes must be serving the demand of subset V_S. If the demand $q(V_S)$ is smaller than the vehicle capacity π, this still implies that $b(V_S) \geq \lceil q(V_S)/\pi \rceil = 1$ vehicle should be visiting that subset, in which case constraints (4.34) are no different to the cutset constraints (4.12) for the TSP. We now show how this formulation can be applied to the CVRP described in Example 4.7.

Example 4.9

We first calculate the number of vehicles to use in the CVRP instance in Example 4.7 with five customers. In this case, both the solution of the bin packing problem and the lower bound procedure yield the same value of $b(V_C) \geq \lceil 9/4 \rceil = 3$ as the minimum number of vehicles needed to find a feasible solution to the problem. We therefore use $|K| = 3$. The objective function of the problem is as follows:

Minimise

$$9.1x_{01} + 12.4x_{02} + 23.3x_{03} + 10.3x_{04} + 19x_{05}$$
$$+ 8.8x_{10} + 13.3x_{12} + 16.5x_{13} + 10.8x_{14} + 20.3x_{15}$$
$$+ 12.8x_{20} + 11.8x_{21} + 25.4x_{23} + 23.1x_{24} + 7.5x_{25}$$
$$+ 25.4x_{30} + 17.1x_{31} + 25x_{32} + 20.8x_{34} + 26.4x_{35}$$
$$+ 9.4x_{40} + 10.1x_{41} + 21.8x_{42} + 19.3x_{43} + 28.5x_{45}$$
$$+ 19.2x_{50} + 20.3x_{51} + 7.6x_{52} + 26.7x_{53} + 29.5x_{54}$$

The optimisation is subject to the following set of constraints. First, two constraints are needed to model the number of vehicles leaving and returning to the depot, as shown in the following:

$$x_{01} + x_{02} + x_{03} + x_{04} + x_{05} = 3$$
$$x_{10} + x_{20} + x_{30} + x_{40} + x_{50} = 3$$

Next, we present the assignment constraints for each customer node $V_C = \{1, 2, 3, 4, 5\}$ written in the same way as in the TSP:

$$x_{10} + x_{12} + x_{13} + x_{14} + x_{15} = 1$$
$$x_{20} + x_{21} + x_{23} + x_{24} + x_{25} = 1$$
$$x_{30} + x_{31} + x_{32} + x_{34} + x_{35} = 1$$
$$x_{40} + x_{41} + x_{42} + x_{43} + x_{45} = 1$$
$$x_{50} + x_{51} + x_{52} + x_{53} + x_{54} = 1$$
$$x_{01} + x_{21} + x_{31} + x_{41} + x_{51} = 1$$
$$x_{02} + x_{12} + x_{32} + x_{42} + x_{52} = 1$$
$$x_{03} + x_{13} + x_{23} + x_{43} + x_{53} = 1$$
$$x_{04} + x_{14} + x_{24} + x_{34} + x_{54} = 1$$
$$x_{05} + x_{15} + x_{25} + x_{35} + x_{45} = 1$$

In order to model tour feasibility, at most $2^5 = 32$ cutset inequalities would be needed, each corresponding to a subset of the customer set. In what follows, we only present some of these inequalities.

For subset $V_S = \{2\}, b(V_S) = 1$:

$$x_{20} + x_{21} + x_{23} + x_{24} + x_{25} \geq 1$$

(already implied by the assignment constraints)

For subset $V_S = \{1, 2\}, b(V_S) = 1$:

$$x_{10} + x_{13} + x_{14} + x_{15} + x_{20} + x_{23} + x_{24} + x_{25} \geq 1$$

For subset $V_S = \{1, 2, 3\}, b(V_S) = 2$:

$$x_{10} + x_{14} + x_{15} + x_{20} + x_{24} + x_{25} + x_{30} + x_{34} + x_{35} \geq 2$$

For subset $V_S = \{2, 3, 4, 5\}, b(V_S) = 3$:

$$x_{20} + x_{21} + x_{30} + x_{31} + x_{40} + x_{41} + x_{50} + x_{51} \geq 3.$$

The last set of constraints needed in the formulation model the integrality of the decision variables.

$$x_{ij} \in \{0, 1\} \quad \forall i, j = 0, \ldots, 5, i \neq j.$$

As this example illustrates, the exponential size of the formulation due to the use of the cutset inequalities necessitates specialised solution techniques for the solution of this formulation. However, as with the TSP, compact formulations are available to formulate CVRP and related problems, as will be shown here.

4.4.1.2 Compact inequalities based on node-ordering for tour feasibility

The first type of inequalities that can be used for the CVRP for preventing tours that are either disconnected from the depot or infeasible with respect to capacity constraints are direct extensions of constraints (4.13) which were introduced for the TSP and are presented as follows:

$$u_i - u_j + \pi x_{ij} + (\pi - q_i - q_j)x_{ji} \leq \pi - q_j \quad \forall i,j \in V_C, i \neq j \tag{4.35}$$

$$q_i \leq u_i \leq \pi \qquad \forall i \in V_C. \tag{4.36}$$

In this formulation, the auxiliary variable u_i corresponds to the load information on the vehicle immediately after having visited customer node i, assuming that the vehicles pick up q_i units from this node. In particular, if $x_{ij} = 1$ for two distinct customer nodes $i,j \in V_C, i \neq j$, then constraint (4.35), written for that customer pair, implies $u_j = u_i + q_j$, where the load u_j carried on the vehicle immediately after visiting node j is calculated as the sum of any load already carried by the vehicle after having left node i, and the demand q_j of customer j. Constraints (4.36) simply state that the load on leaving any customer node should be at least the demand of that customer, and at most the capacity of the vehicle. The case of deliveries can be modelled in a similar way, with the interpretation that u_i variables would, in this case, denote the additional empty space made available in the vehicle after having visited customer i.

4.4.1.3 Compact inequalities based on arc-ordering for tour feasibility

An alternative compact formulation for the CVRP can be obtained by extending the single-commodity flow formulation that was presented for the ATSP by (4.18) and (4.19) in the following way:

$$\sum_{j \in V:(i,j) \in A} f_{ij} - \sum_{j \in V:(i,j) \in A} f_{ji} = q_i \quad \forall i \in V_C \tag{4.37}$$

$$0 \leq f_{ij} \leq \pi x_{ij} \quad \forall (i,j) \in A, \tag{4.38}$$

where variables f_{ij} indicate the actual amount of goods carried on a vehicle when it is travelling from node i to node j. It can be seen that constraints

(4.37) imply an *increasing* amount of flow, as would be the case in pick up or a collection. In other words, these constraints indicate that when a vehicle leaves a (customer) $i \in V_C$ on its way to any other customer node $j \in V_C \backslash \{i\}$ or the depot, it will have picked up q_i units of commodity. This is the reason why constraints (4.37) are only written for customer nodes, although it is possible to write the constraints for all nodes $i \in V$ (including the depot) but where the demand of the depot is set as $q_0 = 0$.

The second set of constraints (4.38) is complementary to the first and has a dual purpose:

1. No flow is allowed on an arc $(i, j) \in A$ if that arc is not being traversed, that is, $f_{ij} > 0$ if $x_{ij} = 1$, and $f_{ij} = 0$ if $x_{ij} = 0$.
2. The amount of flow on any arc is constrained by the capacity π of the vehicle.

It is clear that constraints (4.38) are a linear way of modelling the logical implication between the x and the f variables in exactly the same fashion as in traditional 'big-M' constraints, and the value π is the minimum value that the 'big-M' coefficient can be allowed to take. It is possible to further tighten these constraints for all arcs between customer pairs in the following way:

$$q_i \le f_{ij} \le (\pi - q_j) x_{ij} \quad \forall i, j \in V_C, i \ne j, \tag{4.39}$$

which has an intuitive explanation that, as a vehicle travels from one customer $i \in V_C$ to another $j \in V_C \backslash \{i\}$, the flow f_{ij} has to be at least as large as q_i given that the vehicle will have picked up the demand for node i. Similarly, if $x_{ij} = 1$, which implies that, a vehicle is travelling to node j, needs to have sufficient space available to accommodate the demand q_j of the next customer on the route.

4.4.2 Three-index formulations

All the formulations presented earlier for the CVRP were based on two-index binary variables that provided information for each arc of the graph. If vehicle-specific information needs to be incorporated and presented by such variables, then it becomes necessary to extend the variables to include three indices. In this case, an explicit arc-vehicle-based binary variable z_{ij}^k would be used, which is equal to 1 if arc $(i, j) \in A$ is traversed by vehicle $k \in K$, and 0 otherwise. Such a formulation would be particularly useful when modelling heterogeneous vehicle fleets, where each vehicle $k \in K$ has a capacity π^k or a travel cost c_{ij}^k for traversing arc $(i, j) \in A$. Using the new variables, a three-index formulation for the CVRP can be written as follows:

Minimise $\displaystyle\sum_{(i,j)\in A}\sum_{k\in K} c_{ij}^k z_{ij}^k$

subject to

$$\sum_{j:(0,j)\in A} z_{0j}^k = 1 \qquad \forall k \in K \tag{4.40}$$

$$\sum_{k\in K}\sum_{j:(i,j)\in A} z_{ij}^k = \sum_{k\in K}\sum_{j:(j,i)\in A} z_{ji}^k = 1 \qquad \forall i \in V_C \tag{4.41}$$

$$\sum_{i:(i,p)\in A} z_{ip}^k - \sum_{p:(p,j)\in A} z_{pj}^k = 0 \qquad \forall p \in V, k \in K \tag{4.42}$$

$$\sum_{i\in N_C} q_i \sum_{j\in N:(i,j)\in A} z_{ij}^k \leq \pi^k \qquad \forall k \in K \tag{4.43}$$

$$\text{Tour feasibility} \tag{4.44}$$

$$z_{ij}^k \in \{0,1\} \quad \forall (i,j) \in A, k \in K. \tag{4.45}$$

The structure of this formulation has similarities with that of the one using the two-index variables, in particular constraints (4.40) ensure that all vehicles are used and constraints (4.41) require that each customer is visited exactly once by any vehicle. Unlike the two-index formulation, the three-index variables here carry the information on the particular vehicle traversing on each arc, which means that a vehicle k moving along an arc from any node i to a particular node p should be the same vehicle k leaving node p to any other node j. This requirement is modelled using constraints (4.42). The capacity limitations on each vehicle are represented using constraints (4.43).

The tour feasibility can be formulated in a number of ways, either by using an exponential or a compact set of inequalities described earlier in this chapter. In the following, we show how the node-ordering based compact inequalities (4.35) and (4.36) can be extended for use within the three-index formulation:

$$u_i^k - u_j^k + \pi z_{ij}^k \leq \pi - q_j \quad \forall i,j \in V_C : (i,j) \in A \tag{4.46}$$

$$q_i \leq u_i^k \leq \pi \qquad \forall i \in V_C, \tag{4.47}$$

where continuous variables u_i^k show the position of node $i \in V_C$ within the tour of vehicle $k \in K$. It is easy to show that these two sets of inequalities forbid subtours from forming within the customer nodes.

The three-index formulation is sometimes modelled by using an additional binary variable y_i^k equal to 1 if vehicle $k \in K$ serves customer $i \in N_C$, and 0 otherwise, which is not essential but useful in that it results in a slightly simpler representation as follows:

$$\text{Minimise} \sum_{(i,j) \in A} c_{ij}^k z_{ij}^k$$

subject to

$$\sum_{j:(i,j) \in A} z_{ij}^k = \sum_{j:(j,i) \in A} z_{ji}^k = y_i^k \qquad \forall i \in V_C$$

$$\sum_{i \in V_C} q_i y_i^k < \pi^k \qquad \forall k \in K$$

$$\sum_{i \in V_C} y_i^k = 1 \qquad \forall k \in K$$

Tour feasibility

$$z_{ij}^k \in \{0, 1\} \quad \forall (i, j) \in A, k \in K.$$

In this case, the tour feasibility constraints can be imposed through the following constraints, written for all $V_S \subseteq V_C$, and $V_S \neq \emptyset$,

$$\sum_{(i,j) \in A, i \in V_S, j \notin V_S} z_{ij}^k \geq y_i^k,$$

of which there will be exponentially many.

4.4.3 Set partitioning formulations

Both the two and three-index formulations presented earlier in this chapter use binary variables that only hold *partial* information on the routes needed to arrive at a feasible solution for the CVRP. In other words, each binary variable relates to an arc in the solution and prescribes whether a particular arc will be part of an optimal solution. The constraints are then used to piece this information together so that the routes are feasible with respect to connectivity, capacity and any other additional constraint that one may wish to impose. This is one of the reasons as to why such formulations are rather lengthy in the number of constraints needed. Although intuitive to understand and easy to use with modern off-the-shelf integer programming solvers, they are not necessarily efficient in the use of the number of variables. This is particularly evident in the three-index formulation, the size of which grows with the number of arcs *and* vehicles. For example, a three-index assignment-based formulation for an asymmetric CVRP instance with 25 customers and 5 vehicles will have 3000 binary variables alone.

A different way to think about formulating the CVRP is to associate a variable not with a part of a potential route (such as x_{ij} for each arc), but with a whole route, such that all information needed for and all constraints imposed on the route are captured by such a variable. For this purpose, a set R of all possible routes for the problem will need to be defined and can be generated prior to formulating the problem. It will be easy to calculate the cost c_r of each route $r \in R$. Furthermore, each route $r \in R$ would consist of at least one customer, which can be denoted by a binary parameter a_{ir} that is equal to 1 if customer $i \in V_C$ is in route r, and 0 otherwise. The problem would then simply consist of choosing a subset of routes from within the set R. To this end, it would suffice to define a binary variable λ_r equal to 1 if route $r \in R$ is selected, and 0 otherwise, using which a new formulation for the CVRP can be written as follows:

$$\text{Minimise} \sum_{r \in R} c_r \lambda_r \tag{4.48}$$

subject to

$$\sum_{r \in R} a_{ir} \lambda_r = 1, \qquad \forall i \in V_C \tag{4.49}$$

$$\sum_{r \in R} \lambda_r = |K| \tag{4.50}$$

$$\lambda_r \in \{0, 1\} \quad \forall r \in R. \tag{4.51}$$

The objective function (4.48) simply minimises the total cost of the routes that are selected. Constraints (4.49) indicate that the routes selected should be such that no customer $i \in V_C$ is left out. Finally, constraints (4.50) stipulate that exactly $|K|$ routes should be selected, although this constraint can be left out if there are no requirements on the number of vehicles used. The model (4.48) through (4.51) is often referred to as a *set partitioning* formulation of the CVRP with the reasoning that customers are *partitioned* into the set of routes chosen.

Example 4.10

We return to the vehicle routing problem instance in Examples 4.7 and 4.9, for which $|K| = 3$ identical vehicles of capacity $\pi = 4$ were required to meet the demands expressed as $(q_1, q_2, q_3, q_4, q_5) = (2, 1, 2, 3, 3)$. Bearing in mind the capacity restrictions on the vehicles, the set R of routes, each element of which is feasible with respect to the capacity constraint, can be expressed as follows:

$$R = \{r_1, \ldots, r_{15}\}$$
$$= \{(0,1,0), (0,2,0), (0,3,0), (0,4,0), (0,5,0),$$
$$(0,1,2,0), (0,1,3,0),$$
$$(0,2,1,0), (0,2,3,0), (0,2,4,0), (0,2,5,0),$$
$$(0,3,1,0), (0,3,2,0),$$
$$(0,4,2,0),$$
$$(0,5,2,0)\}.$$

Note that setting $\pi = 4$ has resulted in 15 feasible routes for this instance. Increasing the capacity to $\pi = 5$ would have rendered other routes such as $(0,1,2,3,0)$ or $(0,2,1,3,0)$ feasible. On the other hand, decreasing the capacity down to $\pi = 3$ would have reduced the size of set R even further, as routes such as $(0,1,3,0)$ with a total capacity requirement equal to 4 would have been removed from the set.

Using this notation, a set partitioning formulation for this instance can be written as follows:

$$\text{Minimise } 17.9\lambda_1 + 25.2\lambda_2 + 48.7\lambda_3 + 19.7\lambda_4 + 38.2\lambda_5$$
$$35.2\lambda_6 + 51\lambda_7 + 33\lambda_8 + 63.2\lambda_9 + 44.9\lambda_{10}$$
$$39.4\lambda_{11} + 49.2\lambda_{12} + 61.1\lambda_{13} + 44.9\lambda_{14} + 39.4\lambda_{15}$$

subject to

$$\lambda_1 + \lambda_6 + \lambda_7 + \lambda_8 + \lambda_{12} = 1$$
$$\lambda_2 + \lambda_6 + \lambda_8 + \lambda_9 + \lambda_{10} + \lambda_{11} + \lambda_{13} + \lambda_{14} + \lambda_{15} = 1$$
$$\lambda_3 + \lambda_7 + \lambda_9 + \lambda_{12} + \lambda_{13} = 1$$
$$\lambda_4 + \lambda_{10} + \lambda_{14} = 1$$
$$\lambda_5 + \lambda_{11} + \lambda_{15} = 1$$
$$\sum_{r=1,\ldots,15} \lambda_r = 3$$
$$\lambda_1, \lambda_2, \ldots, \lambda_{15} \in \{0,1\}.$$

The optimal solution λ^* of this formulation has $\lambda_4^* = \lambda_{11}^* = \lambda_{12}^* = 1$ and $\lambda_r^* = 0$ for $r \in R \backslash \{r_1, r_{11}, r_{12}\}$, indicating that the tours $(0,4,0)$, $(0,2,5,0)$ and $(0,3,1,0)$ are in the optimal tour, with a total cost equal to 108.3 miles.

The simplicity of the set covering CVRP formulation is attractive but the challenge lies in producing the set R, which potentially has an exponential number of elements, that is as many as the number of feasible solutions to the problem. For this reason, such formulations are often preferred for problems that are tightly constrained, such as the CVRP with time window constraints, as will be explained in the following section, so that the size of R is manageable.

The size of the instance considered in Example 4.10 is such that it was possible to generate the set R explicitly. For larger instances, this is not practical. Instead, R is initialised with only a (possibly empty) subset $R' \subset R$, and R' is iteratively expanded with the most promising routes $r \in R \backslash R'$ at each iteration. Such an approach is based on principles of *column generation* embedded within a *branch-and-bound* algorithm. The details of such algorithms are beyond the remit of this book, but interested readers can consult the relevant references provided at the end of the chapter.

4.5 Vehicle routing with time windows

Time window requirements are generally defined individually for each customer $i \in V_C$ and in relation to the start time of service at this customer, where service may be either collection (pick up) or delivery. Time windows are specified in the form $[a_i, b_i]$, where $b_i \geq a_i + s_i$, where a_i is the earliest time that a delivery can be made, b_i is the latest time that service can start and s_i is the service time. The VRPTW is a restricted VRP which includes the additional constraint that service at each customer must commence within the interval $[a_i, b_i]$. As was mentioned earlier in the chapter, early arrivals to customers are generally permitted, however this will be at the expense of the driver waiting between the time of arrival and time a_i. One other assumption is that idle waits *after* service has been completed are not allowed. In other words, the driver leaves a customer immediately upon completion of service.

In order to be able to model the time-related aspects of the VRPTW, we introduce the parameter $t_{ij} \geq 0$ for each $(i, j) \in A$ that indicates the travel time from node i to node j defining the arc. As with cost, the travel times can be symmetric or asymmetric. Travel times are often proportional to the distance, but not always, and are generally assumed to be fixed, unless speed is a decision variable (see Chapter 6).

Although time windows are only attributed to customers, it is sometimes convenient to define a time window $[a_0, b_0]$ for the depot. When the travel times satisfy the triangular inequality, the problem parameters should be such that

$$a_0 \leq \min_{i \in V_C}\{b_i - t_{0i}\}$$

$$b_0 \geq \max_{i \in V_C}\{\max\{a_0 + t_{0i}, a_i\} + s_i + t_{i0}\}$$

for the problem to have a feasible solution. It should also be noted that any arc $(i, j) \in A$ for which $a_i + s_i + t_{ij} > b_j$ is infeasible and will therefore never appear in a feasible solution to the problem, and for which reason it can be removed from graph G.

The VRPTW can be modelled using any of the formulations shown earlier, but with additional constraints to reflect the time window requirements. In the rest of this section, we will work with the two-index formulation shown by (4.27) through (4.33), to which the following constraints will be added:

$$a_i \leq v_i \leq b_i \qquad \forall i, j \in V_C, i \neq j \qquad (4.52)$$

$$v_i - v_j + s_i + t_{ij} \leq M_{ij}(1 - x_{ij}) \quad \forall i, j \in V_C, i \neq j. \qquad (4.53)$$

In constraints (4.53), the auxiliary variable $v_i \geq 0$ calculates the *service start* time to node $i \in V_C$, where $M_{ij} = \max\{0, b_i + s_i + t_{ij} - a_j\}$. Constraints (4.52) simply enforce the earliest and latest times at which service can commence. To see how constraints (4.53) work for a pair $(i \in V_C, j \in V_C \backslash \{i\})$ of nodes, consider the following two cases:

1. There is no direct travel from node i to node j, in which case $x_{ij} = 0$, and the constraints read $v_i - v_j \leq b_i - a_j$. Given that $v_i \leq b_i$ and $v_j \geq a_j$ in the light of constraints (4.52), constraints (4.53) will not be binding.
2. There is direct travel from node i to node j, which implies $x_{ij} = 1$, and in which case constraints read $v_j \geq v_i + s_i + t_{ij}$, meaning that the time v_j at which service can start at node j is not earlier than the sum of service start time v_i at node i, the service time s_i at this node and the time it takes to travel from node i to node j.

The following example shows the way that constraints (4.53) and (4.52) operate.

Example 4.11

Figure 4.13 shows a vehicle route, departing from the depot shown by node 0, and visiting, in the given order, customers 1, 2 and 3 to deliver equipment, which also requires installation and testing of the equipment at each location. The time windows in which customers can accept the

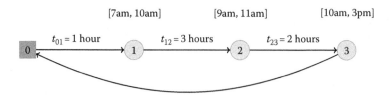

Figure 4.13 A feasible vehicle route under the time window requirements.

deliveries are also shown in the figure, next to each customer. Similarly, the travel time between successive customers is shown next to the corresponding arc. It is assumed, for the purposes of this example, that each delivery takes 1 hour, including the time to off-load the equipment, installation, testing and completion of the relevant paperwork. Constraints (4.53) and (4.52) written for the arcs that appear in the route take the following form:

$$v_1 - v_2 + 1 + 3 \leq 5(1 - x_{12})$$
$$v_2 - v_3 + 1 + 2 \leq 4(1 - x_{23})$$
$$7 \leq v_1 \leq 10$$
$$9 \leq v_2 \leq 11$$
$$10 \leq v_3 \leq 15,$$

where $M_{12} = \max\{0, 10 + 1 + 3 - 9\} = 5$ and $M_{23} = \max\{0, 11 + 1 + 2 - 10\} = 4$. In the calculation of the service start times, one can always adjust the departure time at the depot so that there is no idle wait at the first node. For this particular instance, it can be assumed that the service start time at node is $v_1^* = 7$, indicating that delivery at this node will be completed by 8am, and at which time the driver will continue the route towards customer 2. The arrival time at node 2 will be 11am, which coincides with the upper end of the time window for this node, but is still feasible as it falls within the interval [9:00, 11:00]. Therefore, $v_2^* = 11$. Delivery at customer 2 will be completed at 12 noon, following which the driver will continue on to customer 3, with a planned arrival at 2pm, which will also be the time at which delivery will commence at this node ($v_3^* = 14$), which is within the requested time window and is therefore feasible.

Constraints (4.53) operate in a very similar way to (4.35) which ensure capacity restrictions and can be extended for use within the three-index formulations. The interesting aspect is that constraints (4.53) also act as subtour elimination constraints, as they forbid any tours that are disconnected from the depot.

As with the CVRP, it is possible to formulate the VRPTW using a set covering formulation using a formulation that has the same structure as (4.48) through (4.51) but differs in the way that the feasible set Ω of routes is chosen. In particular, any route $r \in \Omega$ for the VRPTW will need to satisfy the time window restrictions.

4.5.1 Soft time windows

For a VRPTW instance to be feasible, time windows must be such that there exists at least one set of feasible routes that do not violate the related constraints. In such cases, the constraints are said to be *hard*. Finding feasibility under such restrictions may not always be possible, however, in particular for instances where time windows are narrow for each customer. If the customers are willing to accommodate any (small) deviations from the guaranteed slots by the service provider, then the problem can be modelled using *soft* time windows instead, where the desire is still for the service to commence within the time window $[a_i, b_i]$ for each customer $i \in V_C$. If this is not possible, then early or late deliveries would be allowed, but this will be at the expense of some penalty. The penalty is a measure of the loss in service quality, is normally a monetary value and will be assumed to increase proportionally with any deviations from the time windows.

A formulation of the VRP with soft time windows is similar to that of the VRPTW, with the exception of the way in which the objective function is written, as shown here:

$$\text{Minimise} \sum_{(i,j) \in A} c_{ij} x_{ij}$$

$$+ \beta_E \sum_{i \in V_C} \max\{0, v_i - b_i\} + \beta_L \sum_{i \in V_C} \max\{0, a_i - v_i\} \qquad (4.54)$$

subject to

$$\sum_{j \in V_C:(i,j) \in A} x_{0j} = |K|$$

$$\sum_{i \in V_C:(i,j) \in A} x_{i0} = |K|$$

$$\sum_{j \in V:(i,j) \in A} x_{ij} = 1 \qquad \forall i \in V_C$$

$$\sum_{i \in V:(i,j) \in A} x_{ij} = 1 \qquad \forall j \in V_C$$

$$v_i - v_j + s_i + t_{ij} \leq M_{ij}(1 - x_{ij}) \quad \forall i, j \in V_C, i \neq j$$

$$x_{ij} \in \{0, 1\} \qquad \forall (i, j) \in A.$$

In the objective function (4.54), the parameters β_E and β_L denote the unit cost of early or late deliveries, respectively. Although the objective function is non-linear in the way it is stated, it is straightforward to linearise using auxiliary variables.

4.6 Distance-constrained vehicle routing

The problem of finding vehicle routes with additional constraints imposed to limit the length of any route is widely known as the DVRP. Here, the word 'length' can mean the actual physical distance of the route, in which case the problem is to find routes that do not exceed a certain upper bound τ. Length can also be defined as any other measure on which a constraint should be imposed, such as the ones listed here:

- Length could be the total *time* required to traverse the route, in which case τ will correspond to the maximum time that a driver is allowed to be on duty for traversing this particular route.
- Length can be defined as the total *number* of customers on a route, which is particularly relevant where significant time is spent for service at each customer (e.g. bulk deliveries). In this case, τ can be specified as an upper bound on the number of customers that can be served on a route, so as to prevent a driver from performing too many pick up or delivery operations.
- In case where the fleet includes vehicles running on alternative sources of energy, in particular battery, length can be defined as the amount of energy consumption, in which case τ will be the maximum amount of energy that the battery on the vehicle will provide, without needing to be recharged.

In what follows, we will work with the basic problem where length is physical distance, and where d_{ij} defines the distance between nodes i and j on the arc $(i, j) \in A$. The DVRP can be formulated in a similar way to the CVRP. For illustrative purposes, we will once again use the formulation (4.27) through (4.33) and impose the following additional inequalities that model *both* distance restrictions and subtour elimination:

$$y_i - y_j + Mx_{ij} \leq M - d_{ij} \qquad \forall i \in V_C, j \in V_C, i \neq j \qquad (4.55)$$

$$y_i \geq \bar{d}_{0i} + (d_{0i} - \bar{d}_{0i})x_{0i} \qquad \forall i \in V_C \qquad (4.56)$$

$$y_i \leq \tau - \bar{d}_{i0} - (d_{i0} - \bar{d}_{i0})x_{i0} \quad \forall i \in V_C, \qquad (4.57)$$

where $y_i \geq 0$ is an auxiliary variable denoting the length of the path from the depot until node $i \in V_C$. In these constraints, $\bar{d}_{0i}$ denotes the length of the shortest path from the depot to node i. Similarly, $\bar{d}_{i0}$ is the length of the shortest path from node i to the depot. Finally, M is a sufficiently large constant to render the first set of inequalities redundant for when $x_{ij} = 0$, for which any value such that $M \geq \max_{(i,j) \in A}\{\tau - \bar{d}_{0i} - \bar{d}_{i0} + d_{ij}\}$ can be used.

The reader will notice that constraints (4.55) through (4.57) are similar to (4.35) that model the CVRP and (4.53) used to formulate the VRPTW and indeed operate in exactly the same fashion. In particular, constraint (4.55), written for a customer node pair $(i \in V_C, j \in V_C \setminus \{i\})$, defines the relationship between the route length variables y_i and y_j and ensures that the length of the path from the depot to node j is at least d_{ij} units more than the length of the path from the depot to node i on a given route where $x_{ij} = 1$. Constraints (4.56) set a lower bound on the length of the path from the depot to any node $i \in V_C$, which is equal to the length $\bar{d}_{0i}$ of the shortest path if i is not the first node on the route (i.e. $x_{0i} = 0$), but one that changes to d_{0i} should i be the first node on the route (i.e. $x_{0i} = 1$). Constraints (4.57) operate in the same fashion but for the last customer node on a given route. These constraints are polynomial in number, and for which liftings have been proposed.

An alternative way to model route length is to use a single-commodity flow formulation, using variables $f_{ij} \geq 0$ that denote the length of the route from the depot until node $j \in V_C$ for which node i is the immediate predecessor. These constraints are shown here:

$$\sum_{j \in V \setminus \{i\}} f_{ij} - \sum_{j \in V \setminus \{i\}} f_{ji} = \sum_{j \in V \setminus \{i\}} d_{ij} x_{ij} \quad \forall i \in V_C \tag{4.58}$$

$$f_{0i} = d_{0i} x_{0i} \qquad \forall i \in V_C \tag{4.59}$$

$$f_{ij} \leq \tau x_{ij} \qquad \forall (i, j) \in A. \tag{4.60}$$

In the flow formulation, constraints (4.58) are written for each customer node $i \in V_C$ and indicate that the distance from i to any other node $j \in V_C \setminus \{i\}$ should be equal to the difference between the distance from the depot to node i and the distance from the depot to node j. The interesting aspect of these constraints is that distance is modelled as a commodity, and tour lengths are held in the values of the flow of this commodity. The second set of constraints (4.59) initialises the flow values for the first node appearing on each route. Finally, constraints (4.60) are the actual route length constraints that limit the value of any flow in the network to the value τ.

4.7 Uncertainty in vehicle routing problems

The vehicle routing problems that have so far been considered in this chapter assumed that all input data, such as customer demands, travel times or service times, are known with certainty. However, as discussed in Section 4.1, this might not always be the case. The demand of an individual customer, for example, may exhibit some variability. Travel times between customers might depend on traffic and congestion. In most cases, such variability can be captured by analysing historical data, either in the form of probability distributions (either discrete or continuous) or through (a finite set of) scenarios each with a certain probability of occurrence. In this case, one needs to solve a *stochastic vehicle routing problem* by taking into account the random data.

If the actual values that uncertain data will eventually take were known to the planner, at the time of planning, then the problem is no different to the deterministic vehicle routing problems discussed previously. In stochastic vehicle routing problems, however, true values of the random variables will often become known after the plans have been made, and more likely during the execution of the routes. This is also referred to as a *realisation* of the random variables. In case of stochastic vehicle routing problems with collection where the demand is random, for example, the drivers will not know the actual amount of goods to collect from each customer until they arrive at the locations. Random travel times, on the other hand, may be affected by weather conditions (e.g. rainy, foggy or snowy day) and can change during the course of the route. Delays may be encountered due to unexpected congestion (e.g. accidents) or delays as a result of unannounced road works (e.g. repairs), which would not necessarily be known at the time of planning.

In most cases, vehicle dispatching decisions will need to be made *prior* to the routes being executed. Such decisions include the fleet mix, route of each of the vehicle, and the schedule which prescribes the times at which customers will be visited and when service will be provided. Once the decisions have been made, it is assumed that they cannot be changed due to the operational difficulties involved. This is particularly relevant when vehicles of particular size or capacity have already been acquired for the routing (e.g. rented), or when vehicles are already loaded with goods in line with the customers that each vehicle will need to serve on its route.

In its basic form, the stochastic vehicle routing problem consists of finding a solution described by a set of routes in the presence of random data. There are at least two types of approaches to formulate and solve such problems, namely *recourse* models and *probabilistic* (or *chance*) constrained models. These will be explained further in the following sections.

4.7.1 Recourse models

The need to identify and implement a single solution for a stochastic vehicle routing problem might mean that some constraints of the problem, such as those relevant to capacity, time windows or distance, are violated under different realisations of the random data. In other words, the solution might be infeasible under particular scenarios. The point at which there is an infeasibility is said to be a *route failure*. Such failures often mean deterioration of service level or quality. In this case, *recourse* actions need to be taken to remedy the infeasibility, which will invariably have a *penalty cost* associated with them. Here are some examples of recourse actions and penalties:

- In CVRPs where the demand is a random variable, also known as the *vehicle routing problem with stochastic demand*, a route is said to be infeasible if the total demand served on that route turns out to be larger than the vehicle capacity. This arises when a customer on that route has a demand that is larger than planned for, at which point the route is said to *fail*. If the problem only concerns deliveries, a failure will indicate that the vehicle does not hold a sufficient amount of goods to deliver to the relevant customer. If the problem only concerns collections, then the vehicle will not have sufficient capacity available to pick up the amount of goods at the point of failure. In either case, a possible recourse action is for the vehicle to make a return trip to the depot, either to pick up the additional load needed for delivery at the point of failure or to off-load the excess amount of goods to make capacity available for collection at the point of failure. The penalty cost is generally a linear function of the distance between the location of the customer and the depot.
- In the case of the *vehicle routing problem with stochastic travel times*, drivers should not drive more than a preset amount of time that corresponds to their normal working hours. However, in the light of the stochastic data, the actual duration of the route may turn out to be longer than allowed. In this case, one possible recourse action for drivers is to continue the service on the route, but at the expense of a penalty which would correspond to the overtime cost paid to drivers, proportional to the amount of time worked over normal hours.
- When the customers are present with a given probability, meaning that some demand points may not actually require service, the resulting problem is known as the *vehicle routing problem with stochastic customers*. In this case, the solution will assume that all customers are present, and the recourse action for those that are absent is simply to skip such customers. There is normally no penalty associated with this recourse action.

Stochastic vehicle routing problems with recourse seek to find a solution that minimises an expected cost that includes the actual cost of operations (e.g. cost per unit of distance) as well as any penalties arising from infeasibility or route failures. The following section shows the application of a recourse model for a vehicle routing problem where the travel times are random variables.

4.7.1.1 Vehicle routing problems with stochastic travel times

The underlying problem we consider here is the DVRP that is described in Section 4.6, where the length of a route is defined by the time required to traverse it, which should not be longer than τ units which correspond to the maximum normal driving time. We now assume that the travel times t_{ij} from one node $i \in V$ to the other $j \in V \backslash \{i\}$ are uncertain. The realisations of the random variables are described by a finite set $\Xi = \{1, 2, \ldots, \xi, \ldots\}$ of scenarios, where each scenario $\xi \in \Xi$ prescribes a matrix $t^\xi = \{t_{ij}^\xi\}$ of travel times and has a probability p_ξ of occurrence. An overtime cost Δ is charged for per unit of time that drivers spend over their maximum normal time τ. The problem is to find a solution consisting of a set of routes with a minimum cost of traversing the routes and of the expected overtime cost.

There are several ways to formulate this problem. The one we present here is an adaptation of the three-index formulation presented in Section 4.4.2 using the variables defined therein. For the stochastic problem considered here, we define an additional continuous variable ρ_k^ξ that shows the amount of overtime (if any) that drivers spend on route $k \in K$ under the realisation of scenario $\xi \in \Xi$. The formulation is presented as follows:

$$\text{Minimise} \quad \sum_{(i,j)\in A}\sum_{k\in K}c_{ij}z_{ij}^k + \sum_{\xi\in\Xi}p_\xi\sum_{k\in K}\Delta\rho_k^\xi \tag{4.61}$$

subject to

$$\sum_{j:(0,j)\in A} z_{0j}^k = 1 \qquad \forall k \in K$$

$$\sum_{k\in K:(i,j)\in A}\sum z_{ij}^k = \sum_{k\in K:(j,i)\in A}\sum z_{ji}^k = 1 \qquad \forall i \in V_C$$

$$\sum_{i:(i,p)\in A} z_{ip}^k - \sum_{p:(p,j)\in A} z_{pj}^k = 0 \qquad \forall p \in V, k \in K$$

$$\sum_{i\in V_S}\sum_{j\in V_S:(i,j)\in A} z_{ij}^k \leq |V_S| - 1, \quad V_S \subseteq V_C, |V_S| \geq 2, k \in K \tag{4.62}$$

$$\sum_{(i,j)\in A} t_{ij}^\xi z_{ij}^k \leq \tau + \rho_k^\xi \qquad \forall \xi \in \Xi, k \in K \tag{4.63}$$

$$z_{ij}^k \in \{0, 1\} \qquad \forall(i,j) \in A, k \in K.$$

Most of the constraints that appear in this formulation were described in Section 4.4.2. One difference is in the objective function, namely (4.61), where the second term denotes the expected cost of the overtime on each route over all the scenarios. Constraints (4.62) are those described earlier for subtour elimination, although it is possible to use alternative representations for this purpose, such as (4.12) or (4.13). Constraints (4.63) are new, and calculate the overtime ρ_k^ξ for each route $k \in K$ separately, where the term on the left is the total time spent on route k using the travel times in scenario $\xi \in \Xi$. The following illustrative example shows the use of this formulation on a small-scale instance.

Example 4.12

Consider the instance shown in Figure 4.14 with one depot shown by D and four customers $\{1, 2, 3, 4\}$, each of which requires a parcel delivery to be made in a morning shift. The parcel carrier has two vehicles, both of which are to be used in the deliveries. The vehicles have sufficiently large space to carry the types of parcels to be delivered, for which reason it can be assumed that there are no restrictions on vehicle capacity. The plans of the vehicle routes are made the day before the vehicles depart in the morning of a given day and cannot be modified later due to practical reasons.

Drivers should not drive longer than 4 hours in the morning shift and need to be back at the depot for their break within 4 hours of departure. Using historical data, it has been possible to construct two scenarios, ξ_1 and ξ_2 of equal likelihood, prescribing the time it will take to travel from one node of the network to the other. These times are indicated in Figure 4.14 in the form $(t_{ij}^{\xi_1}, t_{ij}^{\xi_2})$ for each link (i, j) shown in the network. For example, in the first scenario, it will take 2 hours to get from the depot to customer 4, whereas the travel time for the same link will be 1 hour in the second scenario. The links in the network are bi-directional, and the travel times specified on each link are the same for

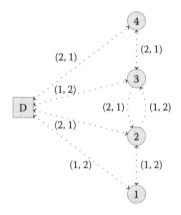

Figure 4.14 A vehicle routing problem instance with $|\Xi| = 2$ scenarios.

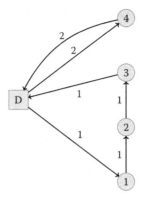

Figure 4.15 An optimal solution for scenario ξ_1.

both directions. We assume, for the sake of simplicity, that all links in the network have the same distance and the same operational cost per unit distance. The objective is therefore to minimise the expected amount of overtime.

It is easy to see that an optimal solution for the first scenario ξ_1 would be as shown in Figure 4.15, consisting of two routes (D, 1, 2, 3, D) and (D, 4, D), where there is no overtime involved for either driver. However, if uncertain data reveal, at the point of departure, that scenario ξ_2 will be realised, then the route (D, 1, 2, 3) will require a total driving time of 8 hours, implying 4 hours of overtime for the driver. A similar result holds for the solution in Figure 4.16, which shows two routes (D, 1, D) and (D, 4, 3, 2, D), both of which require no overtime under scenario ξ_2, but the latter route has an overtime equal to 4 hours under scenario ξ_1. Given that $p_{\xi_1} = p_{\xi_2} = 0.5$, either solution has an expected overtime equal to $4 \times 0.5 = 2$ hours.

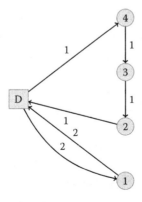

Figure 4.16 An optimal solution for scenario ξ_2.

The question as to whether there exists another solution that is different to the two already identified, and with a lower expected amount of overtime, can be answered by applying the formulation shown earlier. The objective function in this case will read as follows:

$$\text{Minimise } 0.5 \left(\rho_1^{\xi_1} + \rho_1^{\xi_1} \right) + 0.5 \left(\rho_2^{\xi_2} + \rho_2^{\xi_2} \right)$$

which minimises the expected overtime. The following are used to ensure that each customer is visited exactly once.

$$z_{D1}^1 + z_{D2}^1 + z_{D3}^1 + z_{D4}^1 = 1$$

$$z_{D1}^2 + z_{D2}^2 + z_{D3}^2 + z_{D4}^2 = 1$$

$$z_{1D}^1 + z_{12}^1 + z_{1D}^2 + z_{12}^2 = 1$$

$$z_{2D}^1 + z_{21}^1 + z_{23}^1 + z_{2D}^2 + z_{21}^2 + z_{23}^2 = 1$$

$$z_{3D}^1 + z_{32}^1 + z_{34}^1 + z_{3D}^2 + z_{32}^2 + z_{34}^2 = 1$$

$$z_{4D}^1 + z_{43}^1 + z_{4D}^2 + z_{43}^2 = 1$$

$$z_{D1}^1 + z_{21}^1 + z_{D1}^2 + z_{21}^2 = 1$$

$$z_{D2}^1 + z_{12}^1 + z_{32}^1 + z_{D2}^2 + z_{12}^2 + z_{32}^2 = 1$$

$$z_{D3}^1 + z_{23}^1 + z_{43}^1 + z_{D3}^2 + z_{23}^2 + z_{43}^2 = 1$$

$$z_{D4}^1 + z_{34}^1 + z_{D4}^2 + z_{34}^2 = 1.$$

The following constraints model route continuity of vehicles:

$$z_{D1}^1 + z_{21}^1 - z_{1D}^1 - z_{12}^1 = 0$$

$$z_{D1}^2 + z_{21}^2 - z_{1D}^2 - z_{12}^2 = 0$$

$$z_{D2}^1 + z_{12}^1 + z_{32}^1 - z_{2D}^1 - z_{21}^1 - z_{23}^1 = 0$$

$$z_{D2}^2 + z_{12}^2 + z_{32}^2 - z_{2D}^2 - z_{21}^2 - z_{23}^2 = 0$$

$$z_{D3}^1 + z_{23}^1 + z_{43}^1 - z_{3D}^1 - z_{32}^1 - z_{34}^1 = 0$$

$$z_{D3}^2 + z_{23}^2 + z_{43}^2 - z_{3D}^2 - z_{32}^2 - z_{34}^2 = 0$$

$$z_{D4}^1 + z_{34}^1 - z_{4D}^1 - z_{43}^1 = 0$$

$$z_{D4}^2 + z_{34}^2 - z_{4D}^2 - z_{43}^2 = 0.$$

The next set of constraints are used to prevent any subtours from forming within customer nodes, written for vehicle $k = 1$:

$$z_{12}^1 + z_{21}^1 \leq 1$$

$$z_{23}^1 + z_{32}^1 \leq 1$$

$$z_{34}^1 + z_{43}^1 \leq 1.$$

Following are subtour elimination constraints written for vehicle $k = 2$:

$$z_{12}^2 + z_{21}^2 \leq 1$$

$$z_{23}^2 + z_{32}^2 \leq 1$$

$$z_{34}^2 + z_{43}^2 \leq 1.$$

The final set of constraints relates to the calculation of overtime for scenario $\xi = 1$:

$$z_{D1}^1 + 2z_{D2}^1 + z_{D3}^1 + 2z_{D4}^1 + z_{12}^1 + z_{23}^1 + 2z_{34}^1$$

$$+z_{1D}^1 + 2z_{2D}^1 + z_{3D}^1 + 2z_{4D}^1 + z_{21}^1 + 2z_{32}^1 + 2z_{43}^1 \leq 4 + \rho_1^{\xi_1}$$

$$z_{D1}^2 + 2z_{D2}^2 + z_{D3}^2 + 2z_{D4}^2 + z_{12}^2 + z_{23}^2 + 2z_{34}^2$$

$$+z_{1D}^2 + 2z_{2D}^2 + z_{3D}^2 + 2z_{4D}^2 + z_{21}^2 + 2z_{32}^2 + 2z_{43}^2 \leq 4 + \rho_2^{\xi_1}.$$

A similar set of constraints are written for scenario $\xi = 2$ as follows:

$$2z_{D1}^1 + z_{D2}^1 + 2z_{D3}^1 + z_{D4}^1 + 2z_{12}^1 + 2z_{23}^1 + z_{34}^1$$

$$+2z_{1D}^1 + z_{2D}^1 + 2z_{3D}^1 + z_{4D}^1 + 2z_{21}^1 + z_{32}^1 + z_{43}^1 \leq 4 + \rho_1^{\xi_2}$$

$$2z_{D1}^2 + z_{D2}^2 + 2z_{D3}^2 + z_{D4}^2 + 2z_{12}^2 + 2z_{23}^2 + z_{34}^2$$

$$+2z_{1D}^2 + z_{2D}^2 + 2z_{3D}^2 + z_{4D}^2 + 2z_{21}^2 + z_{32}^2 + z_{43}^2 \leq 4 + \rho_2^{\xi_2}.$$

The final set of constraints enforces the integrality of the decision variables z_{ij}^k for all $(i, j) \in A$ and $k = 1, 2$, which will not be explicitly written here.

The solution of this formulation yields an optimal solution that is depicted in Figure 4.17, which includes two routes. The total driving time on route (D, 2, 1, D) is 4 hours under scenario ξ_1 and 5 hours under scenario ξ_2. Similarly, the total driving time on route (D, 3, 4, D) is 5 hours under scenario ξ_1 and 4 hours under scenario ξ_2. The expected amount of overtime is therefore $1 \times 0.5 + 1 \times 0.5 = 1$ hour, an improvement on the deterministic solutions presented in Figures 4.15 and 4.16.

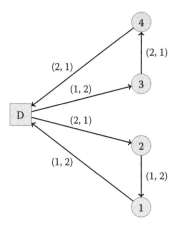

Figure 4.17 An optimal solution of the stochastic programming formulation.

4.7.2 Chance-constrained models

One interesting aspect of the recourse models is the implicit assumption that there is a cost of violating a constraint under a given scenario and that this cost can be quantified. In Example 4.12, the cost of violating the driving-time constraint was the amount of overtime itself and the cost of overtime was explicitly incorporated into the formulation. However, the cost of violating a constraint may not always be possible or well defined, particularly when dealing with intangible consequences of plans made under uncertain data, such as loss of customer goodwill. In this case, one can resort to using what is called *chance-constrained programming*, in which *chance constraints* are used to model the probability of satisfying a set of deterministic constraints. We show how these constraints can be applied for the vehicle routing problem with stochastic demands.

Let the demand q_i of each customer $i \in V_C$ be a random variable with mean $\bar{q}_i$ and a standard deviation ψ_i. It is assumed that the demands are independent. A route failure occurs if the total demand served on any route $k \in \{1, \dots, K\}$ is more than the vehicle capacity π. All other notations are as defined in the previous section. A chance-constrained formulation of the problem will seek to identify a set of routes where the maximum allowable probability of a route failure is $\hat{\alpha}$, as represented by the following set of constraints:

$$P\left(\sum_{(i,j) \in A} q_i x_{ij}^k \leq \pi \right) \geq 1 - \hat{\alpha}. \tag{4.64}$$

Constraints (4.64) seek to find a set of routes where the probability of failure is at most $\hat{\alpha}$ and do not take into account the extent to which the solutions yield failures that are below this probability.

As can be seen from (4.64), chance constraints are not easy to use in their natural form in solving such problems. However, under a number of assumptions, they can be transformed into deterministic equivalent forms, as described by Stewart and Golden (1983) which we detail here. Given that the demands are independent, we can denote the mean of demand on route k by $\Gamma_k = \sum_{(i,j)\in A} \bar{q}_i x_{ij}^k$ and the standard deviation by $\Psi_k = \sum_{(i,j)\in A} \psi_i x_{ij}^k$. If the distribution of $\left(\sum_{(i,j)\in A} q_i x_{ij}^k - \Gamma_k\right)/\Psi_k$ is the same as that of $(q_i - \bar{q}_i)/\psi_i$, which holds for distributions such as normal, Poisson and binomial, then the following constraints can be used in place of (4.64):

$$\Gamma_k + \chi\Psi_k \leq \pi, \tag{4.65}$$

or, equivalently,

$$\sum_{(i,j)\in A} \bar{q}_i x_{ij}^k + \chi\left(\sum_{(i,j)\in A} \psi_i x_{ij}^k\right)^{1/2} \leq \pi, \tag{4.66}$$

where χ is a constant such that the following constraint holds:

$$P\left(\left(\sum_{(i,j)\in A} q_i x_{ij}^k - \Gamma_k\right)\Big/\Psi_k \leq \chi\right) \geq 1 - \hat{\alpha}.$$

Constraints (4.66), although non-linear, are deterministic versions of the chance constraints (4.64), and it would be possible to solve formulations involving such expressions using non-linear programming solvers. There are at least two different ways in which they can be converted into linear constraints, as described here:

1. If the demands q_i, $i \in V_S$ follow a normal distribution for a subset $V_S \subseteq V_C$ of customers, or in case of the number $|V_S|$ of customers being sufficiently large so that the distribution of $\sum_{i\in S} \bar{q}_i$ can be approximated by a normal distribution, the vehicle routing problem with stochastic demands can be solved using a deterministic formulation, such as (4.27) through (4.33), where the tour feasibility is modelled by the following modification of the cutset inequalities (4.34):

$$\sum_{i\in V_S, j\notin V_S} x_{ij} \geq \left\lceil \frac{z_{\hat{\alpha}}(\sum_{i\in V_S} \psi_i)^{1/2} + \sum_{i\in V_S} \bar{q}_i}{\pi} \right\rceil, \tag{4.67}$$

where $z_{\hat{\alpha}}$ is the $1 - \hat{\alpha}$ quantile of the standard normal variable.

2. As Stewart and Golden (1983) observe, further simplifications can be made if ψ_i^2 is a constant multiple of $\bar{q}_i$ as $\psi_i^2 = \kappa\bar{q}_i$, which holds for distributions such as Poisson, binomial, negative binomial and gamma. In this case, constraints (4.66) are reduced to the following:

$$\Gamma_k + \tau(\kappa\Gamma_k)^{1/2} \leq \pi, \tag{4.68}$$

or, equivalently,

$$\Gamma_k \leq \bar{\pi} = (2\pi + \chi^2\kappa - (\chi^4\kappa^2 + 4\pi\chi^2\kappa)^{1/2})/2, \tag{4.69}$$

where the implication is that the vehicle routing problem with stochastic demands where the vehicle capacity is π can be solved as a deterministic vehicle routing problem with vehicle capacity equal to $\bar{\pi}$. Stewart and Golden (1983) note that for larger size instances of the problem, the demand on the route may be approximated by using the normal distribution, in which case the value of χ in constraint (4.66) is the appropriate standard normal variable.

4.8 Dynamic vehicle routing problems

Vehicle routing problems in which all input data, such as customer locations, demands and travel and service times, are sufficiently known in advance and can be solved prior to executing the routes are generally referred to as *static* problems. In practice, however, the information available to the planners is subject to change, or evolves with time, during the execution of the routes, even if they were known with certainty at the time of designing the routes or executing the plans. If the changes in the input data are relatively small, the static plans can still be carried out, particularly if they were made to accommodate such minor changes as in the case of stochastic vehicle routing. If, however, the degree of dynamism in the data is high, then a different approach is needed to solve the resulting *dynamic* vehicle routing problem. Most of such problems consider customer requests as the main source of dynamism, which we will also assume here.

Dynamic vehicle routing problems include two types of requests, one which are known, namely *advance* requests, and the other that arrive over time, known as *immediate* requests. Advance requests are those that are already planned for and for which routes are constructed. Immediate requests are those that are accommodated during the execution phase, to the best extent possible, which might be through inserting them into the routes already planned. Alternatively, immediate requests can be rejected as they might otherwise lead to infeasible routes. The number n_i of immediate requests, combined with the number n_a of advance requests, makes up the total number n_t of requests in the system.

Before discussing the most suitable approaches to use for dynamic vehicle routing problems, it is important to be able to quantify the dynamism in such problems, through what is known as the *degree of dynamism*, which can be calculated as follows:

$$\frac{n_i}{n_t},$$

and it is indicative of the information received in real time in addition to the information already available for the overall system. However, such a simple measure does not differentiate between the different distribution of immediate requests over time. For this purpose, an extended measure, namely the *effective degree of dynamism*, has been proposed to take the arrival times of request into account. Given a planning horizon of length Ω, and the arrival times v_i for all immediate requests $i = 1, \ldots, n_i$ in the system, the effective degree of dynamism is calculated as follows:

$$\frac{\sum_{i=1,\ldots,n_i} v_i/\Omega}{n_t},$$

which can be interpreted as a normalised average of the arrival times of immediate requests. The following example illustrates how the two measures can be applied in practice.

Example 4.13

A parcel carrier has planned vehicle routes for use in making deliveries to nine customers, starting at 8am on a given day. The carrier also accepts immediate requests until 6pm on the same day. This implies that the length of the planning horizon is $\Omega = 10$ hours. If a total of three requests come in throughout the day, then the degree of dynamism is $3/12 = 1/4$.

On the other hand, if three requests come in at 9, 10 and 11am, respectively, then the effective degree of dynamism in the system is calculated as

$$\frac{(1/10) + (2/10) + (3/10)}{12} = 0.05.$$

Finally, if three requests were to come in at 12 noon, 2pm and 4pm, respectively, then the effective degree of dynamism in the system would be

$$\frac{(4/10) + (6/10) + (8/10)}{12} = 0.15.$$

Dynamic vehicle routing problems are generally solved by initially considering advance requests alone, which produces a *base plan*. Such plans can be made by solving the static version of the problem using approaches shown earlier in the chapter. Immediate requests, if accepted, are then inserted into the existing routes, which could be done using at least three different ways, some of which are described here.

1. *Reoptimisation* (re)solves either the entire problem, or part of it, to deal with the changing nature of the input data. There are two ways of implementing a reoptimisation approach:

    a. Instant reoptimisation is performed whenever there is a new (immediate) request arriving into the system. This approach is suitable for systems that do not exhibit a high degree of dynamism.

    b. Periodic reoptimisation is run at discrete points in time, ideally equally spaced over the implementation period, where the time interval between two time points is called a decision epoch. The solution obtained at the start of a decision epoch would be the one to be implemented over that epoch, and no further changes to the solution within the epoch would be allowed. This strategy is preferred over instant reoptimisation for systems that have a high degree of dynamism, particularly if instant reoptimisation is relatively difficult to perform or implement.

2. *Update policies*: Reoptimisation is generally suitable for dynamic vehicle problems where the optimisation itself is not very costly (in terms of, e.g., computational running time). If the optimisation itself is difficult or requires significant computational time, an alternative strategy is to locally update the existing plans using predefined policies. This is normally done by inserting new requests into the existing plans. Various criteria, such as those related to geographical and temporal proximity, latency, response time or one which provides minimal disruption, can be used to perform the insertion. Generally, however, those that result in the shortest detour across all the vehicles are preferred. It is important that these rules are such that their implementation does not require complex methods and demand additional computational time, as they would need to run fast and ideally provide good-quality solutions.

3. *Anticipatory strategies*: Repositioning the vehicles or using waiting strategies in anticipation of future requests is another way of coping with immediate requests. When implementing a dynamic VRP solution, a vehicle, at any point in time, is serving a customer, moving to another customer or waiting idle at some location. After serving all advance requests, two different types of decisions can be considered. The first is for the vehicle to wait at the location where it served last, as opposed to going back to the depot, in anticipation of future requests. Alternatively, the vehicle might be moved to a location that is expected to be closer to future request(s) than any other location in the network.

For the latter strategy to be effective, however, some knowledge of future requests is needed to be able to identify promising locations. This can be extracted from historical data in the form of stochastic information.

The following example illustrates how some of the concepts mentioned earlier can be applied in practice.

Example 4.14

A carrier with one vehicle located at a depot shown by D in Figure 4.18 has received an advance order which consists of picking up five pallets from node A within the time window $[1, 3]$ and delivering it to node C within the time window $[1, 10]$, where the times specified are in hourly units. The vehicle can carry up to ten pallets.

The travel time between each pair of locations is also shown next to each node of the symmetric graph depicted in Figure 4.18. The driver has to return back to the depot after completing the job. Pick up and delivery times at each location are negligible. If the carrier is concerned with minimising the total travel time of the driver, then an optimal route for this problem instance is given by D–A–C–D, with a total driving time equal to 12 hours, assuming that the driver starts driving at time 0 and reaches node A at time $t = 3$ and node C at time $t = 8$.

Additional demands for pickups and deliveries arrive continuously and randomly throughout the day and are received by the carrier. Assume that a new order is received at time $t = 3$, which consists of picking up a further five pallets from node B (see Figure 4.19) within the

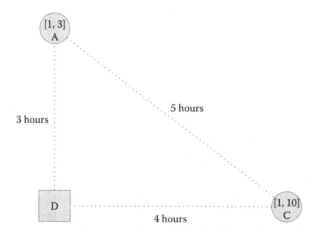

Figure 4.18 A dynamic vehicle routing problem instance.

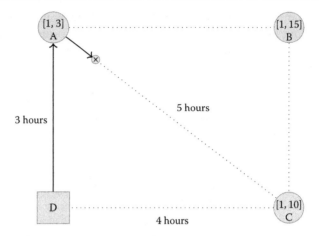

Figure 4.19 Position of the vehicle at the time of receiving the new order.

time window $[1, 15]$ and bringing it back to the depot at anytime during the day. Consider, now, the two following scenarios:

1. In the first scenario, the carrier uses periodic reoptimisation using decision epochs of length equal to two time units. At time $t = 2$, for example the vehicle will still be en route from the depot to node A, and as the new order will not yet have been received at this time, the original route D–A–C–D will remain intact. At time $t = 4$, the vehicle will have picked up the first load from node A and will be on the way to node C, as shown by the '×' in Figure 4.19. At this point, the carrier will need to decide whether to accept the new order or not, based on considerations such as feasibility, real-time capacity availability or other limitations. If, for example, the time window at node B was $[1, 10]$, then the carrier would have to reject the new order, as the vehicle would not be able to make it to node B in time, given that it is already destined to node C. Similarly, if the total driving time was restricted to be within 12 hours, then the new order would once again have to be rejected. If, on the other hand, there are no such considerations preventing the carrier to accept the new order, then the resulting route would be such as the one shown in Figure 4.20, which implies arriving at node B at time 11, and with a total driving time of 16 hours.

2. We now consider the scenario where the carrier uses instant reoptimisation, which means that a new route can be constructed at time $t = 3$ of receiving the new order, which is when the vehicle is at location A and has already picked up the first load. In this case, the vehicle accepts the new request, moves on to node B to pick up the load, then travels to node C to deliver the first five pallets and continues on to the depot to deliver the remaining five pallets. The total

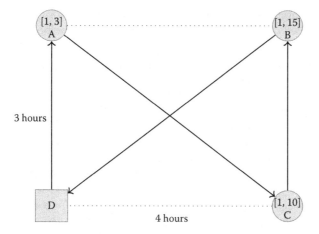

Figure 4.20 Optimal route after vehicle accepts new order o_2 at time $t = 4$ (periodic reoptimisation).

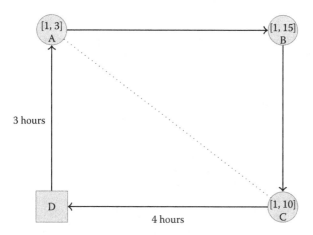

Figure 4.21 Optimal route after vehicle accepts new order o_2 at time $t = 3$ (instant reoptimisation).

driving time in this case is 14 hours, which is a saving of 2 hours over the solution of the first scenario, indicative of the savings of using an instant reoptimisation approach (Figure 4.21).

4.9 Practical application: Charity collection

The case study is described by Erdoğan et al. (2015), which is a vehicle routing problem arising in charity collection. Unwanted goods such as clothes, shoes, books or other material can be donated to charitable organisations (charity, in short) by either taking them directly to shops, or to donation banks provided by charities. The banks are bins that are located in places of easy access, such as supermarkets, public car parks or recycling centres. The charity is responsible for emptying donation banks as they fill, which results in a complex routing problem.

Such a problem is faced by a major charity operating in the United Kingdom with a network of approximately 650 shops selling new and used goods, and about 1300 donation banks across the nation. The charity operates a fleet of vehicles to collect goods from banks and shops, for delivery to processing centres, as well as to move donated items between shops where they would be most likely to be resold. Each item is associated with a sell value, and the collection problem faced by the charity is to find routes for the vehicle fleet that visit each location (bank or shop) no more than once, with the objective of maximising the net profit, computed as the difference of the value of collected donations and the travel cost incurred. All vehicles start from and return to a single depot. However, this must be done under a number of practical restrictions listed as follows:

- Each shop has an opening and closing time, which are translated into time windows and vary based on the parking restrictions. Some of the shops may only be visited at certain times of the day relating to whether access during shop opening hours is required, prohibited or not desired. If a vehicle arrives to a shop before the opening time, it is permitted to wait outside until the shop opens. In contrast, banks can be visited at any time of the day, subject to the working hours of drivers. The opening and closing times of the depot are 4.30am and 5pm, respectively, to ensure that all vehicle rounds were no longer than 12.5 hours in duration. In practice, a majority of drivers start their rounds at 4.30am trying to avoid peak traffic as far as possible.
- Each type of vehicle has an associated travel cost per mile and load capacity. The fleet of vehicles consisted of a single van and five lorries for this particular case study, the capacities of which are known through their technical specifications. Two vehicle types were described in the case study, namely a transit van with gross vehicle weight equal to 3.5 tonnes and a carrying capacity of 1.4 tonnes, and a lorry with gross vehicle weight equal to 12 tonnes and a carrying capacity of 6 tonnes. Transport costs were assumed to be £0.93/km (=£1.50/mile) for the van and £1.09/km (=£1.75/mile) for the (five) lorries, based on estimated figures provided by the charity organisation.
- All vehicles are subject to limits on the driving time and the working time, and drivers are required to rest if they exceed a given driving time limit or a given working time limit. EU regulations require the drivers to rest for 45 minutes for every 4.5 hours of accumulated driving. Similarly, the policy of the charity organisation is such that the drivers

should rest for 45 minutes for every 6 hours of continuous work. In conjunction, these rules require the driver to keep track of the working time and driving time accumulated within a day, and to take a break whenever one of them reaches the limit. Vans and lorries are subject to different regulations, where the maximum driving time is 10 hours for the former and 9 hours for the latter. Similarly, the maximum working time for a van driver is 11 hours and for a lorry driver is 12.5 hours.

One particular challenge faced by the charity organisation in this complex routing problem is the uncertainty over the collection amounts in donation banks, as it is difficult to know the exact amount of goods donated in a bank and accumulated in that bank on the day of visit. In reality, exact weights of the goods in a bank will never be known; however, a reasonable assumption is that the weight of goods (kg) available to be collected from a site on any given day is capable of being estimated well. One can overcome this difficulty by estimating the collection amounts through the use of a probability distribution, particularly if historical data suggest that the collection amounts follow a stochastic pattern. For this case study, a truncated normal distribution was used to model the donation amounts, which ranged from 0 to the capacity of the banks. A bank consists of up to four bins, each of which has a 270 kg capacity, and the capacity of a bank is a multiple of the number of bins it contains. The routing problem that arose in the case study is related to a smaller network in a particular region of the United Kingdom, which consisted of 58 donation banks, 75 shops and the depot. The input data needed to formulate the problem included driving distances and driving times between all pairs of the 134 sites, equating to a total of 8911 entries. These data can be obtained from online maps or from commercial global positioning system (GPS) software. The problem can be modelled as a mixed-integer programming formulation, but given the number of nodes and the practical constraints, it is too complex to be solved by an off-the-shelf solver directly. However, the formulation can be used within a heuristic method, a so-called *matheuristic*, which can handle problem instances of this scale. The algorithm was used in practice by the charity organisation over a period of 36 working days between 9 May and 19 July 2013 with routes being computed one day in advance of being implemented. The choice of the algorithm was due to its robust performance with the mandatory shop visits. Historical data were used to estimate daily fill levels at the other banks or where the remote monitoring sensors were not functioning satisfactorily. The use of these data and the algorithm resulted in an overall 28% reduction in the number of bank visits, from 953 to 685 visits over the period, resulting in average distance and time savings of 3.2% and 2.8%, respectively per day, over fixed routes that would normally have been undertaken in the absence of optimised routes. It was not possible to measure the exact monetary value of the collection, but the algorithm yielded improvements not only in distance and time but also in CO_2 emissions of around 464 kg, based on an assumed average emissions rate of 400 g/km and the predominant use of trucks with a carrying capacity of 6 tonnes. In practice, the transport manager of the charity performs manual alterations to the plans made due to daily operating conditions, which increase the complexity of the problem and are difficult to be included in the models or algorithms devised to solve this problem. Such alterations are made due to reasons such as unexpected traffic which affects driving times, additional and urgent requests made by shops or bank managers or last-minute unavailability of a vehicle due to breakdowns or staff being sick.

References and further reading

TSP owes its fame not just due to its practical relevance, but arguably more so to its theoretical importance in the field of combinatorial optimisation. For an excellent introduction to the topic, the reader is referred to the books by Lawler et al. (1985) and Gutin and Punnen (2006) that include a mathematical treatment of the problem and its variations, by Applegate et al. (2011) which has a computational focus and by Cook (2012) for a non-mathematical exposition on the origins of the problem, applications and solution methods.

Since its inception in the work of Dantzig et al. (1954), the vehicle routing problem has been subject of extensive research. The books by Golden et al. (2008) and Toth and Vigo (2014) provide an extensive coverage of the topic, ranging from problem variants (including routing problems with capacitated vehicles, heterogeneous fleet, time window restrictions and stochastic or dynamic input data) and models to (exact and heuristic) algorithms and applications (including small package delivery, health care logistics, routing in container terminals and disaster relief). For further reading, one may consult the review by Laporte (1992) that includes a number of formulations for the CVRP, by Laporte (2009) that surveys algorithmic progress on the problem over a period of 50 years until 2009, by Pillac et al. (2013) and Bektaş et al. (2014) for dynamic vehicle routing problems, by Gendreau et al. (2014) for stochastic vehicle routing problems, by Ghiani et al. (2003) for real-time vehicle routing problems and by Koç et al. (2016b) for heterogeneous vehicle routing problems. A more detailed treatment of dynamic models for within freight transportation is presented by Powell et al. (2007).

For a detailed treatment of the use of chance constraints within the stochastic vehicle routing problem, see Stewart and Golden (1983). Formulations for the DVRP were described as early as in the work of Kulkarni and Bhave (1985), later corrected by Achuthan and Caccetta (1991) and subsequently improved by Naddef (1994). Various formulations for the vehicle routing problem with time windows are described by Kallehauge (2008).

The models presented in this chapter are based on those described in the references above.

Chapter 5

Integrated routing problems

The previous chapters have so far discussed a variety of planning problems arising in the design of a freight distribution network, separately at strategic, tactical and operational levels of planning. Each problem was modelled and solved in isolation within the planning level in which it arises, with no integration considerations with problems arising in other levels. However, there is an inherent interdependency between the decisions made at different levels of planning. To address this interdependency, planners sometimes take a hierarchical approach, whereby strategic problems are solved initially, the outputs of which are then used as inputs to the tactical and then to operational problems. For example, the vehicle routing problem (VRP) introduced in Chapter 4, which is at an operational level of planning, assumed that the strategic decision concerning the location of the depot(s) was already made. Such an approach, however, is likely to result in suboptimal solutions for the system as a whole, as the decisions made at one level will be highly dependent on the quality of the decisions made at other levels. It is for this reason that some integration between problems of different levels would be necessary.

Planning models should ideally incorporate as many decisions as is practically possible to exploit the interdependencies and be solved to arrive at an overall optimum design for the distribution network as a whole. However, such an ambitious treatment gives rise to very complex and challenging planning problems, for which most, but not all, of the resulting models are intractable. Good progress has been made on some integrated routing problems insofar as being able to jointly model and solve problems at two different levels of planning. The subject of this chapter is to present three types of such problems, which all share one common decision problem: vehicle routing. For this reason, they will collectively be referred to as *integrated routing problems* and will exploit the interactions between vehicle routing and three other planning problems, namely those involving location, inventory and production.

We first start with a problem that integrates some of the basic location problems studied in Chapter 2 with those of routing presented in Chapter 4.

5.1 Location-routing problem

The first type of integrated routing problem concerns a joint treatment of facility location and vehicle routing and is broadly known as the *location-routing problem (LRP)*. In particular, given a potential set of depot locations, a set of customers each with a demand for a particular commodity and a set of delivery vehicles, the decisions concern the location of depots and the routing of vehicles from the depots so as to minimise a given cost function.

The LRP generalises the VRP and the facility location problem. In order to illustrate how the two decisions affect each other, we present an example.

Example 5.1

Consider an instance defined on a 3×4 rectangle as shown in Figure 5.1. The network has two potential depot locations shown by D_1 and D_2 and with three customers labelled as A, B and C. We assume that a single vehicle is available for use with sufficient capacity to be able to serve all three customers with the amounts of goods they have requested.

If a depot is located on node D_1, both (D_1, A, B, C, D_1) and (D_1, C, B, A, D_1) will be optimal tours, as shown in Figure 5.1, with a total route length of 14 units. If, on the other hand, a depot is located on node D_2, then the optimal tours are shown in Figure 5.2, namely (D_2, A, B, C, D_2) or (D_2, C, B, A, D_2), each with a length of 12 units. The latter solution is preferable from a distance-minimising perspective so long as the cost of locating a depot at D_1 or D_2 is the same, or if the latter is at most two units more than the former. This example suffices to show the impact that the depot location has on the shape of the resulting tour and on the total cost.

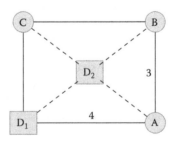

Figure 5.1 Optimal solution shown with solid lines with a total distance of 14 units.

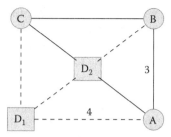

Figure 5.2 Optimal solution shown with solid lines with a total distance of 12 units.

More details on the characteristics of the LRP are presented in the following, most of which are similar to those of the VRP presented in Chapter 4 and the facility location problem presented in Chapter 2. We provide a brief summary here for the sake of completeness:

1. *Customers*: There exists a set of customers (individuals or retailers), each with given locations and demand for a (positive) amount of the product to be delivered. Travelling to and between customers implies a unit routing cost, often expressed as a function of the distance between a given pair of locations. Each customer is generally required to be visited by only one vehicle, which should deliver the demand in full.
2. *Vehicle fleet*: A fleet of vehicles is available with which deliveries are to be made, and the total number of vehicles is often limited to a predefined amount. If there is unlimited fleet, then the use of a vehicle incurs a one-off fixed cost, irrespective of the amount of load carried or the number of customers it serves. If all vehicles in the fleet are uniform with respect to a given set of specifications, often the vehicle capacity, or they are of the same type, then the fleet is said to be *homogeneous*. Otherwise the fleet is *heterogeneous*. As in the VRP, the vehicles might have several restrictions, two of which are detailed here:
   a. Vehicles have capacity limits pertaining to the amount of load that can be carried. Within the context of location-routing problems, they are treated to be one-dimensional, and as such can be defined with respect to volume (in case of liquids or gases), weight or, in rare cases, length.
   b. There might be restrictions on the total length of the tour traversed by a vehicle, imposed either by the available fuel or driver regulations, or both.
3. *Depots*: There exists a discrete set of potential locations to locate depots, each associated with a fixed location or opening cost, and where the goods are supplied from. The fixed costs need to be defined such that

they are comparable with the routing costs. In particular, they need to relate to the same time period in which both problems are solved. As the routing problem is often operational and solved on a short-term (e.g. daily) basis, it is convenient to define an appraisal period for the facilities (e.g. 10 or 20 years) based on which the fixed costs can be amortised.

Depots might have restrictions on their delivery capacities, giving rise to the *capacitated LRP*. Capacities might be defined in terms of loading dock space (in m^2), number of delivery doors, storage space (in m^2 or m^3) or processing (throughput). If the delivery capacities are sufficiently large (i.e. at least as large as the total amount demand to be served for all customers), or not relevant, then the problem is known as the *uncapacitated LRP*.

Each depot can be associated with a subset of vehicles from the vehicle fleet, from where routes should start and to which the vehicles should return.

5.1.1 Formal problem definition

In this section, we will formally define the capacitated LRP with homogeneous vehicles and introduce a formulation that can be used to solve the problem optimally. For reasons of simplicity, we will henceforth simply refer to this problem as the LRP.

The LRP is defined on a graph G with a set V of nodes often partitioned as $V = V_D \cup V_C$, where V_D is the available locations at which depots can be located and V_C is the set of customers. Locating a depot at location $d \in V_D$ incurs a fixed-cost g_d, which is associated with a particular delivery capacity of Π_d. It is assumed that only one type of commodity is to be supplied from the depots and delivered to customers. A set K of vehicles is available and can be distributed across the located depots. Each vehicle has a load capacity π. Each customer $i \in V_C$ will have a positive demand π_i for the commodity.

The problem is to decide on (1) the number and location of new depots and (2) the routes of a set of vehicles, such that each customer is visited by exactly one vehicle and served in full, each vehicle returns to the depot from where it has started the tour, the depot and vehicle constraints on capacity are respected and each route includes only one depot. The objective is to minimise the combined cost of locating depots and vehicle routes.

We will now present two formulations for the LRP, both of which are based on different types of flow. The first one is an exponential-sized formulation and is based on the flow of vehicles. The second one is a compact formulation that is polynomial in the number of constraints and describes the flow of two commodities on the network.

5.1.2 Formulation based on flow

The first formulation we present for the LRP assumes a directed graph G, where there are arcs between every pair of customers and arcs between each potential depot location and the customer set. In other words, the arc set is defined as $A = A_{DC} \cup A_C$, where $A_{DC} = \{(i,j) | i \in V_D, j \in V_C\}$ and $A_C = \{(i,j) | i,j \in V_C, i \neq j\}$. Travelling on each arc $(i,j) \in A$ incurs a cost c_{ij} that either represents the distance between nodes i and j or is a function thereof.

The formulation uses a binary variable x_{ij}^k that is equal to 1 if vehicle $k \in K$ travels on arc $(i,j) \in A$, and 0 otherwise. A second binary variable z_{id} will be equal to 1 if customer $i \in V_C$ is assigned to a depot on location $d \in V_D$, and 0 otherwise. Finally, a third binary variable y_d is used if a depot is located on node $d \in V_D$. Using these variables, a formulation for the LRP can be described as follows:

$$\text{Minimise} \sum_{d \in V_D} g_d y_d + \sum_{k \in K} \sum_{(i,j) \in A} c_{ij} x_{ij}^k \tag{5.1}$$

subject to

$$\sum_{k \in K} \sum_{i \in V::(i,j) \in A} x_{ij}^k = 1 \qquad \forall j \in V_C \tag{5.2}$$

$$\sum_{i \in V} \sum_{j \in V_C:(i,j) \in A} q_i x_{ij}^k \leq \pi \qquad \forall k \in K \tag{5.3}$$

$$\sum_{i \in V_C} q_i z_{id} \leq \Pi_d y_d \qquad \forall d \in V_D \tag{5.4}$$

$$\sum_{d \in V_D} \sum_{i \in V_C} x_{di}^k \leq 1 \qquad \forall k \in K \tag{5.5}$$

$$\sum_{k \in K} \sum_{i \in V_S} \sum_{j \in V \backslash V_S} x_{ij}^k \geq 1 \qquad \forall V_D \subseteq V_S \subset V \tag{5.6}$$

$$\sum_{j \in V:(i,j) \in A} x_{ij}^k - \sum_{j \in V:(i,j) \in A} x_{ji}^k = 0 \qquad \forall k \in K, i \in V \tag{5.7}$$

$$\sum_{h \in V:(i,j) \in A} x_{ih}^k + \sum_{j \in V_C:(d,j) \in A} x_{dj}^k \leq z_{id} + 1 \qquad \forall d \in V_D, i \in V_C, k \in K \tag{5.8}$$

$$y_d \in \{0,1\} \qquad \forall d \in V_D \tag{5.9}$$

$$z_{id} \in \{0,1\} \qquad \forall d \in V_D, i \in V_C \tag{5.10}$$

$$x_{ij}^k \in \{0,1\} \qquad \forall (i,j) \in A, k \in K. \tag{5.11}$$

In this formulation, the first term of the objective function (5.1) is the cost of locating depots and the second term is the cost of routing, the summation of which are minimised. Constraint (5.2) is used to ensure that each customer is visited on the route of only one vehicle. Load capacity constraints on the vehicles are represented by constraints (5.3). Constraints (5.4) model the situation that if a depot is located at node $d \in V_D$, in which case $y_d = 1$, then a delivery capacity of at most Π_d units will be made available for customers allocated to this particular node. Constraints (5.5) stipulate that each route will include at most one depot.

The remaining set of constraints refer to the connectivity requirements of the routes. Constraints (5.6) are the usual subtour elimination constraints, as in the TSP, used to ensure that there exists no routes that visit only customer nodes. Constraints (5.7) are generally called *conservation of flow* and ensure continuity of routes. In particular, they guarantee that if a node i is visited by vehicle k^*, then the same vehicle k^* leaves node i for travel onto the next node on the route. The final set of constraints (5.8) is used to link the assignment variables to the routing variables. In particular, if a vehicle k leaves a depot at node d for any node $j \in V_C$, in which case $x_{dj}^k = 1$, and the same vehicle leaves customer i to visit any other node $h \in V$, in which case $x_{ih}^k = 1$, then the constraints force the assignment variable $z_{id} = 1$ as node i will then have to be assigned to depot d.

This formulation can be extended to deal with cases, for example, where there is a constraint imposing a maximum allowable length τ on the tour each vehicle can traverse, as shown here:

$$\sum_{i \in V} \sum_{j \in V:(i,j) \in A} c_{ij} x_{ij}^k \leq \tau \quad \forall k \in K. \tag{5.12}$$

The formulation shown by (5.1) through (5.11) is of exponential size, due to the combinatorial nature of constraints (5.6) needed to model subtour elimination. It is possible to replace these constraints by the compact constraints for the TSP described in Section 4.2.2.

5.1.3 Formulation based on two commodities

An alternative way to model the LRP is to consider a different formulation that relies on an expansion of the graph and the use of additional variables that describe the flow of two commodities. One of these commodities is the actual amount of goods flowing on the network. The other commodity is used to measure the empty or remaining space in the vehicle. The formulation assumes that there is an unlimited number of vehicles available for use from any depot.

This formulation is described on a symmetric and augmented graph which is defined with respect to edges $\{i, j\}$ as opposed to arcs (i, j). The augmented graph $G' = (V', E')$ uses a copy $V_{D'}$ of the node set V_D representing the depots, where $V' = V_D \cup V_C \cup V_{D'}$ and the set of edges is defined as $E' = E \cup \{\{d', i\} : d' \in V_{D'}, i \in V_C\}$.

The variables needed to formulate the problem are as follows. A continuous variable f_{ij}^d defines the flow on edge $\{i, j\} \in E'$ and a 'reverse' variable f_{ji}^d shows the remaining capacity of the vehicle traversing the same edge, such that $f_{ij}^d + f_{ji}^d = \pi$ if edge $\{i, j\}$ is traversed. This will be introduced as a constraint in the formulation. A binary variable x_{ij}^d is used that is equal to 1 if a vehicle dispatched from facility d travels on edge $\{i, j\} \in E'$, and 0 otherwise. For single-customer routes involving a depot $d \in V_{D'}$ and node $i \in V_C$, a new binary variable η_{di} is used, which equals 1 if edge $\{d, i\}$ is used twice, once from $d \in V_D$ to $i \in V_C$ and once for the way back, and 0 otherwise. We retain the binary variables z_{di} and y_d from the previous formulations.

The so-called *two-commodity* formulation is then presented as follows:

$$\text{Minimize} \sum_{d \in V_D} g_d y_d + \sum_{d \in V_D} \sum_{\{i,j\} \in E'} c_{ij} x_{ij}^d + \sum_{d \in V_D} \sum_{i \in V_C} c_{ij} \eta_{ij} \tag{5.13}$$

subject to

$$\sum_{j \in V' : \{i,j\} \in E', \{j,i\} \in E'} x_{ij}^d + 2\eta_{di} = 2z_{id} \quad \forall i \in V_C, d \in V_D \tag{5.14}$$

$$\sum_{d \in D} z_{id} = 1 \quad \forall i \in V_C \tag{5.15}$$

$$x_{di}^d + \eta_{di} \leq z_{id} \leq y_d \quad \forall i \in V_C, d \in V_D \tag{5.16}$$

$$\sum_{j \in V' : \{j,i\} \in E'} f_{ji}^d - \sum_{j \in V' : \{i,j\} \in E'} f_{ij}^d + 2q_i \eta_{di} = 2q_i z_{id} \quad \forall i \in V_C, d \in V_D \tag{5.17}$$

$$\sum_{i \in V_C} f_{di}^d + \sum_{i \in V_C} q_i \eta_{di} \leq \Pi_d y_d \quad \forall d \in V_D \tag{5.18}$$

$$\sum_{i \in V_C} f_{d'i}^d - \pi \sum_{i \in V_C} f_{di}^d = 0 \quad \forall d' \in V_{D'} \tag{5.19}$$

$$f_{ij}^d + f_{ji}^d = \pi x_{ij}^d \quad \forall \{i, j\} \in E', d \in V_D \tag{5.20}$$

$$f_{ij}^d, f_{ji}^d \geq 0 \quad \forall \{i, j\} \in E', d \in V_D \tag{5.21}$$

$$y_d \in \{0, 1\} \quad \forall d \in V_D \tag{5.22}$$

$$z_{id} \in \{0, 1\} \quad \forall d \in V_D, i \in V_C \tag{5.23}$$

$$x_{ij}^k \in \{0, 1\} \quad \forall \{i, j\} \in E, k \in K \tag{5.24}$$

$$\eta_{di} \in \{0, 1\} \quad \forall d \in V_D, i \in V_C. \tag{5.25}$$

In this formulation, the objective function (5.13) minimises the total cost of routing. Constraints (5.14) link the customer-depot assignment decisions with the routing decisions and guarantee that every customer is visited exactly once. Constraints (5.15) ensure that every customer i is assigned to exactly one depot node d, and if this is the case, then the next set of constraints (5.16) ensures that a depot must be located on node d. They also guarantee that if node i is to be visited by a vehicle that has been dispatched from depot d, either on a single or multiple node tour, then the corresponding assignment variable z_{di} takes the value 1. Constraints (5.17) link the flow variables with the assignment variables. Capacity constraints on the depots are represented by constraints (5.18). Constraints (5.19) show that a total of π units of the second commodity are made available for every vehicle that is based at depot d. Constraints (5.21) define the relationship between the two variables, f_{ij}^d and f_{ji}^d, as was also mentioned earlier. The last four sets of constraints define the nonnegativity and integrality restrictions on the decision variables.

5.2 Inventory-routing problem

The routing problems studied in this book, and in particular VRP, assume that all delivery (or collection) operations are to take place within a single time period. The period is often defined in daily time units, such as in the distribution of fresh milk or newspapers, where the assumption is that customer demands must be met in the exact amount specified, at some point in the day. This means that customers cannot be delivered more or less than the amount requested, which further assumes that each customer implements their own inventory policy that dictates how much of a given commodity to request on each day. In such situations, the supplier (or the vendor) and the shipper, who are often the same entity, are limited in terms of consolidating customer demands and often find themselves in situations where vehicles are not efficiently utilised with respect to load capacity.

One way of improving the situation, both with respect to cost and vehicle utilisation rates and to afford the supplier further flexibility with their delivery operations, is to adopt a so-called *vendor-managed inventory* system, in which the supplier decides not only the design of the delivery routes but also the timing and the amounts of commodities to be delivered to each customer on different routes (i.e. replenishment of inventory for all customers). In such cases, the decisions of inventory and routing need to be considered jointly, giving rise to what is called the class of *inventory-routing problems (IRPs)*. In the IRP, customers are willing for their inventories to be managed by the vendor, involving how much and the frequency with which to be supplied with the required commodity. Customers will normally be restricted by a maximum level of inventory they could hold, often dictated by the holding capacities at their premises (e.g. space available in their warehouse).

A large number of variants in this class of problems can be defined by taking into account the following variations:

1. *Planning horizon*: The planning horizon can either be of finite or infinite length.
2. *Shipment times*: Shipments to a customer can be made either at any point in time during the planning horizon, which is referred to as continuous shipments. Alternatively, a minimum intershipment time τ can be imposed between successive deliveries to a given customer. If the shipments can only take place at successive intervals $k\tau$ where $k \in \{0, 1, 2, \ldots\}$, then shipment times are said to be discrete.
3. *Replenishment policy*: A wide variety of replenishment policies can be implemented in line with the requirements of customers. Some examples are given here:

   a. *Maximum Level (ML)*: Under the ML policy, a customer can be served by any amount of goods such that, when combined with the available inventory at the time of the visit, it will not exceed the customer's maximum inventory level.
   b. *Order-up-to Level (OL)*: The OL policy is a special case of the ML policy in that the amount delivered is equal to the difference between the maximum inventory level allowed at the customer location, and the available inventory at the time of the visit.
   c. *Zero-inventory Level (ZL)*: The ZL policy dictates that a customer is visited when their inventory level drops to 0.

4. *Depletion of inventory*: In the general definition of the IRP, one does not explicitly specify, in a given period, the exact point in time when the goods in the inventory are consumed. The general convention is that deliveries are made before consumption (e.g. early in the day), and the inventory level at the end of a period is calculated after deliveries have been made and the demand for that period has been met. The manner in which inventory is depleted can either be deterministic or stochastic. In the former case, the rate of consumption is a known constant. In the latter case, it is characterised by a probability distribution.

5. *Inventory levels*: A common assumption in an IRP is that inventory levels at customers never drop below 0. In some variants, however, one may allow them to be negative, in which case one of the following two situations occurs:

   a. If the customer is willing to wait for the missing amount, also called a back-order, it will be supplied at a later point in time within the planning horizon, but possibly incurring extra cost.
   b. If the customer is unable to wait for the missing amount, then it is considered lost for good, and the vendor faces a possible loss of customer goodwill, and a cost penalty.

6. *Cost*: The cost is often a combination of routing and inventory costs, where the latter includes those relevant to holding inventory, back-orders and lost-sales.

Example 5.2

We now provide a small IRP instance to serve as an example of how the problem is defined and can be solved. The instance is defined on the same graph as in the previous section, using a 3 × 4 rectangle, where there is a single depot shown by D and three customers denoted by A, B and C, whose daily demands for a particular commodity are 50, 30 and 70, respectively, and who hold no starting inventories. Assume there is an unlimited fleet of homogeneous vehicles each able to carry up to 100 units of the commodity. We assume that the objective is to reduce the total transportation cost, which, for ease of convenience, is assumed to be proportional to the total distance travelled by all the vehicles. We would like to model and solve this problem over a two-day period, with the intention that the resulting solution is repeated every two days over a certain period of time.

We also assume that we wish to solve this problem instance with as few vehicles as possible, in which case at least two vehicles are needed. One feasible solution is shown in Figure 5.3, where one vehicle visits

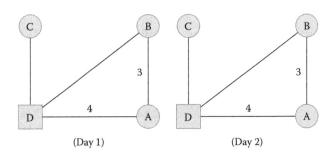

(Day 1)   (Day 2)

Figure 5.3 One solution to the 2-day inventory-routing problem instance.

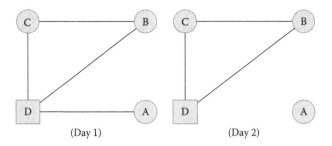

(Day 1)   (Day 2)

Figure 5.4 An alternative solution to the 2-day inventory-routing problem instance.

customer C to deliver 70 units, and another vehicle follows a tour on which customers A and B are served and are delivered 50 and 30 units, respectively. The same routing pattern repeats itself on day 2, and the total transportation cost is 36 units, split equally over the 2 days. In this case, the capacity utilization rates of the two vehicles are 70% and 80%.

It is not difficult to find an alternative solution to this instance where one vehicle visits customers B and C, delivering 30 and 70 units, respectively, and the other vehicle visits customer A alone, delivering a total of 100 units of the commodity, enough to last this customer for 2 days. The solution is shown in Figure 5.4. The solution shows that customer A will not need to be visited on day 2, on which only the tour of the first vehicle to serve customers B and C needs to be repeated. The total cost of this solution is 32 units, a four-unit improvement over the first. If the cost of holding inventory at customer A is at most four units, then the solution shown in Figure 5.4 is preferred to the solution in Figure 5.3 with respect to an objective function comprising transportation and inventory costs. In the latter solution, the capacities on the two vehicles are used with 100% efficiency.

It is also possible to come up with a third solution for the problem instance which uses three vehicles, V1, V2 and V3, where V1 visits customer A on day 1 delivering 100 units, V2 visits customer B on day 1 delivering 60 units, and V3 visits customer C on both days delivering 70 units each time. Although this solution implies a smaller transportation cost of 30 units, it might or might not be preferred over the ones shown in Figures 5.3 and 5.4 depending on the cost of the additional vehicle and the combined holding cost at customers A and B.

5.2.1 Formal problem definition

The IRP can be defined on an undirected graph $G = (V, E)$ with $V = \{0, \ldots, n\}$ as the set of nodes, $V_C = V \backslash \{0\}$ as the set of customers and $E = \{\{i, j\} : i, j \in V, i < j\}$ is the set of edges, where a positive travel cost c_{ij} is associated with each edge $\{i, j\} \in E$. A depot is located on node 0, which serves as the supply point. Each customer $i \in V_C$ has a fixed and known consumption rate q_i^t of a commodity in each time period t of a finite planning period $T = \{1, \ldots, t, \ldots\}$. In each period $t \in T$, there is a total of r^t units of commodity available at the supplier. There exists a set K of vehicles, and vehicle $k \in K$ has a capacity of π^k units. Inventory can be kept at either the supplier or any of the customer nodes $i \in V_C$ up to an amount Π_i, but this incurs a unit holding cost h_i.

The problem is to find, for each period $t \in T$, the delivery amounts to each customer and the associated delivery routes, such that

- Each customer $i \in V_C$ is supplied with a certain amount such that it has at least q_i^t units to meet its demand in period $t \in T$, and after which inventory levels are nonnegative but do not exceed the holding capacity.
- Each customer is visited by at most one vehicle.
- Each vehicle does at most one tour delivering at most π^k units in total, starting and ending at the depot.

The objective is to minimise the total delivery (routing) and holding (inventory) cost. Given that the rates at which inventories are depleted are known at each node, we will refer to this version of the problem as the deterministic IRP.

5.2.2 Exponential-size formulation

A formulation for the deterministic IRP uses four sets of variables, two of which are integers. Of the latter, one set is associated with the routing component of the problem, denoted by x_{ij}^{kt} which is equal to the number of times the edge $\{i, j\}$ is used by vehicle $k \in K$ in period $t \in T$. For edge $\{0, i\}$, this variable can take the values 0, 1 or 2, with the last value indicating a tour in the form $(0, i, 0)$, involving a single customer $i \in V_C$. For all other edges $\{i, j\}$ with $i, j \in V_C$ and $i < j$, the variable is equal to 1 if that edge is used, and equal to 0 if not. Another binary variable y_i^{kt} is equal to 1 if customer $i \in V_C$ is visited by vehicle $k \in K$ in period $t \in T$, and in this case a continuous variable ζ_i^{kt} shows the amount delivered to this customer. Finally, the continuous variables ϕ_i^t show the amount of inventory left at the end of period $t \in T$, and this is *after* the demands have been met at nodes for $i \in V_C$, where ϕ_i^0 denotes the available inventory at node i, which can be zero.

With the use of the four sets of variables as defined earlier, a formulation for the deterministic IRP can be written as follows:

$$\text{Minimise} \sum_{t \in T} \left(\sum_{i \in V_C} h_i \phi_i^t + \sum_{\{i,j\} \in E} \sum_{k \in K} c_{ij} x_{ij}^{kt} \right) \tag{5.26}$$

subject to

$$\phi_1^t - \phi_0^{t-1} - r^t + \sum_{k \in K} \sum_{i \in V_C} \zeta_i^{kt} = 0 \qquad \forall t \in T \qquad (5.27)$$

$$\phi_i^t - \phi_i^{t-1} + q_i^t - \sum_{k \in K} \zeta_i^{kt} = 0 \qquad \forall i \in V_C, t \in T \qquad (5.28)$$

$$\phi_i^{t-1} + \sum_{k \in K} \zeta_i^{kt} - \Pi_i \leq 0 \qquad \forall i \in V_C, t \in T \qquad (5.29)$$

$$\phi_0^t - \Pi_0 \leq 0 \qquad t \in T \qquad (5.30)$$

$$\zeta_i^{kt} - \Pi_i y_i^{kt} \leq 0 \qquad \forall i \in V_C, k \in K, t \in T \qquad (5.31)$$

$$\sum_{i \in V_C} \zeta_i^{kt} - \pi^k y_0^{kt} \leq 0 \qquad k \in K, t \in T \qquad (5.32)$$

$$\sum_{j \in V: \{i,j\} \in E} x_{ij}^{kt} + \sum_{j \in V: \{j,i\} \in E} x_{ji}^{kt} = 2 y_i^{kt} \qquad i \in V, k \in K, t \in T \qquad (5.33)$$

$$\sum_{i,j \in V_S: \{i,j\} \in E} x_{ij}^{kt} - \sum_{i \in V_S} y_i^{kt} + y_m^{kt} \leq 0 \qquad \begin{array}{l} \forall V_S \subseteq V_C, k \in K, \\ t \in T, m \in V_S \end{array} \qquad (5.34)$$

$$\sum_{k \in K} y_i^{kt} \leq 1 \qquad \forall i \in V_C, t \in T \qquad (5.35)$$

$$\phi_i^t \geq 0 \qquad \forall i \in V, t \in T \qquad (5.36)$$

$$\zeta_i^{kt} \geq 0 \qquad \forall i \in V, k \in K, t \in T \qquad (5.37)$$

$$x_{0i}^{kt} \in \{0, 1, 2\} \qquad \forall i \in V_C, k \in K, t \in T \qquad (5.38)$$

$$x_{ij}^{kt} \in \{0, 1\} \qquad \forall i, j \in V_C, k \in K, t \in T \qquad (5.39)$$

$$y_i^{kt} \in \{0, 1\} \qquad \forall i \in V, k \in K, t \in T. \qquad (5.40)$$

In this formulation, constraints (5.27) ensure that the amount of commodity made available by the supplier in period $t \in T$, combined with the inventory I_0^{t-1} from the previous period $t - 1 \in T$, is used to deliver to the customers by all vehicles, and any leftover amount I_0^t is the inventory for period t. In a similar way, constraints (5.28) model the inventory changes for each customer $i \in V_C$ in each period $t \in T$.

Inventory capacity restrictions are modelled through four sets of constraints. The first set, namely constraints (5.29), simply limits the total amount of inventory held at each customer in a given period by the available capacity. In particular, these constraints stipulate that in a period $t \in T$, the existing inventory from the previous period $t - 1 \in T$ combined with the delivery made in that period cannot exceed the available capacity C_i at customer $i \in V_C$. Constraints (5.30) limit the inventory held at the supplier by the available capacity. The following sets of constraints link the deliveries made to customers by the inventory capacity, in particular constraints (5.31) do not allow any deliveries to a customer $i \in V_C$ by a vehicle $k \in K$ in any period $t \in T$ unless a visit has been made, in which case $y_i^{kt} = 1$.

Capacity restrictions on the vehicles are modelled using constraints (5.32), whereby no vehicle $k \in K$ can carry more than its capacity π^k. Constraints (5.33) and (5.34) model the degree constraints and eliminate subtours, where the former set guarantees that each customer is visited by exactly one vehicle in any period and the latter set ensures that there are no tours that are disconnected from the depot. The requirement that each customer is visited by at most one vehicle in each period is modelled with constraints (5.35), although this constraint can be removed if one wants to allow for split deliveries, that is, the case where multiple vehicles visit a customer for supplying the commodity. The remaining set of constraints in the formulation represents the nonnegativity and integrality restrictions on the variables.

5.2.3 Polynomial-size formulation

The formulation presented earlier for the IRP is exponential in size due to the subtour elimination constraints (5.34) which are written for every subset of the node set V_C. It is possible to provide an alternative formulation for the problem, that is polynomial in size, in a similar way as was shown for the TSP in Chapter 4, by considering an asymmetric graph $G = (V, A)$ where A is the set of arcs between every pair of nodes. In this case, variables x_{ij}^{kt} would be redefined to indicate whether a vehicle $k \in K$ in period $t \in T$ travels from node $i \in V$ to node $j \in V \setminus \{j\}$. Auxiliary continuous variables f_i^{kt} would be used to denote the load of vehicle $k \in K$ when it arrives at node $i \in V_C$ in period $t \in T$. In this case, it would suffice to replace constraints (5.33), (5.34), (5.38) and (5.39) in the earlier formulation with the following:

$$\sum_{j \in V} x_{ij}^{kt} = y_i^{kt} \qquad \forall i \in V_C, k \in K, t \in T \tag{5.41}$$

$$\sum_{j \in V} x_{ji}^{kt} = y_i^{kt} \qquad \forall i \in V_C, k \in K, t \in T \tag{5.42}$$

$$f_i^{kt} - f_j^{kt} + \pi^k x_{ij}^{kt} \leq \pi^k - q_j^{kt} \quad \forall i, j \in V_C, i \neq j, k \in K, t \in T \tag{5.43}$$

$$q_i^{kt} \leq f_i^{kt} \leq \pi^k \qquad \forall i \in V_C, i \neq j, k \in K, t \in T \tag{5.44}$$

$$x_{ij}^{kt} \in \{0, 1\} \qquad \forall i, j \in V, k \in K, t \in T \qquad (5.45)$$

$$f_i^{kt} \geq 0 \qquad \forall i \in V, k \in K, t \in T, \qquad (5.46)$$

which would provide for a valid and a polynomial-size formulation of the IRP. Constraints (5.41) and (5.42) ensure that each customer, if visited in a period $t \in T$, is served by a single vehicle. Constraints (5.43) are of the same type used in the compact node-ordering formulation discussed in Chapter 4, and constraints (5.44) restrict the load of each vehicle by its capacity. The last two sets of constraints are the integrality and nonnegativity restrictions on the decision variables.

5.3 Production-routing problem

The IRP that was presented in Section 5.2 assumed that there was a fixed quantity r^t of commodity available at the supplier in period $t \in T$, and that this amount was sufficient to meet the total demand in that period. This assumption is based on the premise that the supplier has already determined the amount of production in each period, by solving a lot-sizing problem, and independent of the inventory and routing decisions. The production, when it takes place, incurs fixed setup and variable unit costs. The supplier also has the option to hold inventory at their premises, when needed. However, solving this problem independently of the routing and deliveries does not guarantee an overall optimal solution for the whole chain of production, distribution, delivery and storage operations, as the following example illustrates.

> **Example 5.3**
>
> Consider the instance given in Example 5.2. In the first solution shown in Figure 5.3, the daily demands are 50, 30 and 70, for customers shown by A, B and C, respectively. The total amount of production required in this case would be 150 units per day, over the 2-day planning period. The solution shown in Figure 5.4 requires 200 units to be produced on day 1, of which 100 units would be delivered to customers B and C and 100 units would be delivered to customer A. In contrast, only 100 units would need to be produced on day 2 to be sent to customers B and C. If the supplier had a production capacity of 150 units a day, then the routing shown in Figure 5.4 would not be feasible, even though it results in lower transportation cost. This is because the total amount of production required on day 1 would be short by 50 units in relation to the total quantity that needs to be delivered.

If the production consideration is integrated into the IRP to jointly optimise production, inventory and routing, this gives way to what is known as the *Production Routing Problem* (PrRP). In this problem, the production and inventory quantities are also decision variables. The IRP is therefore a special case of the PrRP, where the decisions relevant to production have already been made. This observation implies that one can extend the IRP formulation presented in Section 5.2.2 to model the PrRP, using an additional decision variable corresponding to production quantities at the supplier. It should be noted that this formulation uses routing variables x_{ij}^{kt} which include a vehicle index k that explicitly models the route of vehicles. In what follows, we present an alternative formulation which does not use such an index and therefore has a fewer number of binary variables.

5.3.1 Formal problem definition

The PrRP is modelled on a similar undirected graph as the IRP and uses the same notation in terms of the unit inventory costs, transportation costs, demands at customers, vehicle capacities and the maximum inventory levels at each location. The additional parameters are relevant to the unit production cost at the supplier (or the production facility), denoted by $\lambda > 0$ and the fixed production setup cost denoted by $g > 0$. The production capacity available at the supplier is Υ in each period.

5.3.2 Mathematical modelling

Some of the variables used for the PrRP formulation are derived from that of the IRP presented in the previous section, but without the vehicle index, assuming the vehicles are identical each with capacity π. In particular, we define a variable x_{ij}^t which is equal to the number of times the edge $\{i,j\}$ is used in period $t \in T$. For edges incident to the depot, this variable can either be 0, 1 or 2. For all other edges, it is binary. Similarly, the binary variable y_i^t is equal to 1 if customer $i \in V_C$ is visited in period $t \in T$, and in this case a continuous variable d_i^t shows the amount delivered to this customer. If $i = 0$, then variables y_0^t, defined separately for each period $t \in T$, denote the number of vehicles used in that period and are therefore defined as general integer.

The *new* variables specific to the PrRP come in two sets. The first set includes the binary variable y_t that is equal to 1 if there is production in period $t \in T$. The other set models the amount of production performed through the use of the continuous variables v_t, one for each period $t \in T$. A formulation of the PrRP is given as follows:

$$\text{Minimise} \sum_{t \in T} \left(\lambda v_t + g y_t + \sum_{i \in V} h_i \phi_i^t + \sum_{\{i,j\} \in E} c_{ij} x_{ij}^t \right) \tag{5.47}$$

subject to

$$\phi_1^t - \phi_0^{t-1} - p^t + \sum_{i \in V_C} \zeta_i^t = 0 \qquad \forall t \in T \tag{5.48}$$

$$\phi_i^t - \phi_i^{t-1} + q_i^t - \zeta_i^t = 0 \qquad \forall i \in V_C, t \in T \qquad (5.49)$$

$$I_i^{t-1} + \zeta_i^t - C_i \leq 0 \qquad \forall i \in V_C, t \in T \qquad (5.50)$$

$$\zeta_i^t - C_i y_i^t \leq 0 \qquad \forall i \in V_C, k \in K, t \in T \qquad (5.51)$$

$$\sum_{i \in V_C} \zeta_i^t - Q y_1^{kt} \leq 0 \qquad t \in T \qquad (5.52)$$

$$\sum_{j \in V:\{i,j\} \in E} x_{ij}^{kt} + \sum_{j \in V:\{j,i\} \in E} x_{ji}^{kt} = 2y_i^t \qquad i \in V, t \in T \qquad (5.53)$$

$$\pi \sum_{i,j \in V_S:\{i,j\} \in E} x_{ij}^t - \sum_{i \in V_S} \left(\pi y_i^t - \zeta_i^t \right) \leq 0 \qquad \forall V_S \subseteq V_C, |V_S| \geq 2, t \in T$$

$$(5.54)$$

$$y_0^t \leq |K| \qquad t \in T \qquad (5.55)$$

$$0 \leq v^t \leq \Upsilon \qquad t \in T \qquad (5.56)$$

$$\phi_i^t \geq 0 \qquad \forall i \in V, t \in T \qquad (5.57)$$

$$\zeta_i^t \geq 0 \qquad \forall i \in V, k \in K, t \in T \qquad (5.58)$$

$$x_{0i}^t \in \{0,1,2\} \qquad \forall i \in V_C, k \in K, t \in T \qquad (5.59)$$

$$x_{ij}^t \in \{0,1\} \qquad \forall i,j \in V_C, k \in K, t \in T \qquad (5.60)$$

$$y_i^t \in \{0,1\} \qquad \forall i \in V_C, k \in K, t \in T \qquad (5.61)$$

$$y_0^t \in \mathbb{Z} \qquad \forall i \in V_C, k \in K, t \in T. \qquad (5.62)$$

The objective function (5.47) of this formulation minimizes a total cost function comprising fixed and variable costs of production, inventory holding at the production facility and delivery costs to customers. Constraints (5.48) through (5.62) are very similar to the constraints used in the formulation of the IRP. The additional constraints here are (5.56), which limit the production by the available capacity at the supplier in each period. We also point out that the subtour elimination constraints (5.54) in this case are represented in a different way but are used for the same purpose of eliminating tours that are not connected to the production facility.

References and further reading

The literature on the location-routing problem is rich. Formulations and computational comparisons are presented by Albareda-Sambola et al. (2005) and Contardo et al. (2013). Algorithms have been described as early as 1985, in particular see Perl and Daskin (1985) and Laporte et al. (1986, 1988). More recent exact algorithms are described by Baldacci et al. (2011) and Belenguer et al. (2011) for the capacitated location-routing problem. Interested readers are referred to Nagy and Salhi (2007), Prodhon and Prins (2014), Drexl and Schneider (2015) and Albareda-Sambola (2015) for reviews on the location-routing problem, variants and extensions.

The inventory-routing problem has equally been subject of much research. An introduction to the subject can be found in Bertazzi and Speranza (2012). For a review of the literature from the first application of the problem from 1983 to 2012, the readers are referred to the survey by Coelho et al. (2013). Formulations and algorithms have been described by Archetti et al. (2007) and Coelho and Laporte (2014) for the inventory-routing problem, and by Coelho and Laporte (2013a,b) and Coelho et al. (2012) for the variants of the IRP, such as those with multiple products, homogeneous and heterogeneous vehicles and additional consistency features.

The production-routing problem is relatively new and the research on the problem is yet to reach maturity. A review of formulations, solution algorithms and computational results for the problem is provided by Adulyasak et al. (2015b). Algorithms are described by Adulyasak et al. (2012, 2013) for the problem and by Adulyasak et al. (2015a) for the case where there is uncertainty in customer demand.

The models presented in this chapter are based on those described in the references above.

Chapter 6

Green freight distribution

6.1 Introduction

Freight transport is crucial to economic activity, as it allows an increase in the flow of goods globally. Investment in transportation is fundamental for economic growth and development; the movement of goods from one place to another enables the exchange of these goods, stimulating trade and commerce, and thus economic development. However, transport comes at a price having undesirable environmental impacts, which are generally referred to as *externalities*.

As the environmental impacts of freight transport increase with the flow of globally shipped goods, the need to mitigate these is paramount. There are often several objectives that freight transportation strives to meet, which relate to economic, social and environmental aspects. However, it is not always possible to meet all the objectives simultaneously, which sometimes results in a compromise in one or more of the objectives, depending on the trade-offs.

This chapter will describe ways in which some of the externalities listed earlier can be taken into account in developing freight transportation planning models. The consideration of such effects within transportation and logistics planning to help reduce the environmental effects has given rise to a research field now known as *green logistics*, where the aim is to find environmentally friendly ways of designing transportation and logistics activities.

We first briefly describe some of the externalities of transport and their effects on the environment.

6.1.1 Emissions and pollution

Being arguably the most prominent of externalities, emissions are a consequence of use of fuel within the transport sector, such as liquid fuel, most of which is produced from petroleum. Gasoline and diesel are two of the most widely used types of liquid fuels, both of which are composed of many different types of hydrogen and carbon compounds, known as hydrocarbons, in which hydrogen and carbon atoms are chemically bound. Hydrocarbons include, for example, methane (CH_4) and ethane (C_2H_6).

Engines used in a majority of transportation vehicles, and particularly in road transport, produce energy by mixing hydrocarbons with air. Volume-wise, a large fraction of air contains oxygen (O_2) and nitrogen (N_2). The process by which energy is produced by engines is known as *internal combustion*, which, under ideal conditions, should only result in carbon dioxide (CO_2) and water vapour (H_2O). However, the process is incomplete, meaning that not all of the fuel is consumed during internal combustion. As a result, a number of by-products are generated, such as carbon monoxide (CO) as a result of insufficient oxygen, nitrogen oxide (NOx) as a result of nitrogen reacting with oxygen and particulate matter (PM) which are small particles of dust, soot and organic matter suspended in the atmosphere. In addition, small amounts of methane (CH_4) and nitrous oxide (N_2O) are also produced during the combustion process. Furthermore, when sunlight reacts with air that contains hydrocarbons and NOx, it produces another gas called ozone (O_3). In maritime transportation, ships running on diesel engines burn fuel that contains high amounts of sulphur content, as a result of which sulphur oxide as well as nitrogen oxide emissions (also known as NOx and SOx) are significant. The existence of these gases in the atmosphere gives rise to air pollution.

Some of the environmental impacts of emissions and air pollution are listed here:

- *Greenhouse effect*: The various types of gases mentioned earlier, namely H_2O, CO_2, CH_4, N_2O and O_3, are alternatively known as greenhouse gases (GHGs). They trap longwave radiation and prevent it to pass out of the atmosphere, that results in warming of the planet's surface.
- *Acidification*: Pollutants from emissions, in particular NOx and SOx, released into the atmosphere are converted to acids, which fall back as precipitation, eroding soil, limestone and man-made materials.
- *Summer smog*: NOx and hydrocarbons from transport are chemically transformed by sunlight causing thick dense fog, polluting the air and reducing visibility.
- *Depletion of the ozone layer*: The ozone layer is vital for life on Earth, as it absorbs harmful shortwave ultraviolet radiation. The use of ozone-depleting chemicals and chlorofluorocarbons continues to damage the atmosphere.

These changes contribute to the disruptions in the climate, with potentially harmful effects on human health and the environment.

6.1.2 Noise and vibration

Transport gives rise to noise pollution, primarily in urban areas, with undesirable effects on human health. These effects can range from short-term annoyance, sleep disturbance and impaired cognitive functioning to significant long-term physiological problems such as cardiovascular diseases, loss of hearing and mental health problems.

6.1.3 Land and resource consumption

Natural resources are depleted for use in the form of fuel, as well as by extraction of the materials required for building or construction. In addition, the infrastructure itself necessary for transportation activities (roads or highways, railway tracks, ports or terminals, storage or distribution facilities) damages land and local ecosystems.

6.1.4 Toxic effects

Increasing levels of chemicals from transport can negatively affect human health, known as *human toxicity (toxic effects on humans)*, as well as *ecotoxicity (toxic effects on ecosystems)*, where increasing levels of chemicals from transport can negatively affect the environment and the organisms which live within it. One other undesirable consequence is *eutrophication*, where organic or mineral nutrients from transport fuels enter into bodies of fresh water, promoting excess algae growth, drawing oxygen away from other plants or animals, and reduce biodiversity.

6.2 Mitigating measures

Various approaches have been suggested to reduce the environmental impact of freight transport, some of which are described here:

- *Technological* improvements are aimed at developing new technology to improve the environmental performance of freight vehicles, including improved design of vehicles that allow for additional carrying capacity without compromising the fuel or energy efficiency, new vehicles that run on alternative sources of energy (such as electric vehicles or alternative fuels such as biodiesel) and new design or redesign of engines that have better energy efficiency.
- *Strategic or tactical* approaches include changing the so-called *modal split* of freight and transferring freight from energy-intensive modes of transport, such as road, to more environmentally friendly modes of transport, for example rail, coastal shipping, waterways, or use of any

of these in combination with road transport. Tactical approaches also include fleet management involving the selection of the right vehicles to purchase and maintain, as well as regular vehicle maintenance to ensure that they operate at optimum efficiency. In road transportation, for example underinflated tyres, fuel leaks or poor combustion can lead to loss in fuel efficiency of a vehicle (see McKinnon, 2010, for further details).

- *Operational* strategies include reducing the actual numbers of vehicles running, vehicle kilometres and tonne kilometres by increasing load factors (reducing empty or partly loaded running of lorries), utilising new information technology to improving the routing and scheduling of vehicles, consolidating deliveries, sharing loads and pick up deliveries with other companies and improving driver training and behaviour.

It is important that these approaches are implemented to produce win-win solutions, in which the environmental benefits should be at least as much as the gains reaped by the organisations or companies who adopt these practices.

6.3 Quantifying externalities

The prominent externalities of road transportation, including accidents, noise and greenhouse gas emissions, are not always easy to quantify. As GHGs are more prominent among all externalities and are easier to quantify as opposed to others, this section will present models that aim to estimate the total amount of emissions. GHG emissions are generally reported in carbon dioxide equivalent (CDE, CO_{2e} or CO_{2eq}) units, calculated by taking into account the global warming potential of a gas in reference to the amount of CO_2 that would have the same global warming potential over a period of time. For example, that the global warming potential for CH_4 is calculated as 25, with respect to a period of 100 years, implies that one unit of CH_4 is equal to 25 units of CO_2 in terms of emissions.

There are at least two ways to estimate fuel consumption for vehicles, including *on-road measurements* which are based on a real-time collection of emissions data on a running vehicle, and *analytical fuel consumption* (or *emission*) models which estimate fuel consumption based on a variety of vehicle, environment and traffic-related parameters, such as vehicle speed, load and acceleration. These models range from simple to more complex ones, depending on the number of parameters that they take into account to be able to make the estimations. In the following, we present a number of fuel consumption models.

6.3.1 Fuel consumption models

Analytical models for fuel consumption can be broadly classified into three classes, namely (1) emission factor models, (2) average speed models and (3) modal (including instantaneous) models. Emission factor models are the simplest in form and are used at a macroscale level (regional or national emission estimations), particularly when data related to a vehicle's journey are limited. They use an emission factor often expressed per unit of distance. Average speed models are speed-related functions to estimate emissions at a road network level and do not include detailed enough parameters for an analysis at a microscale level. In contrast, modal models operate at a higher level of complexity, specific enough for use at a microscale level and use detailed inputs, such as acceleration and road gradient, which are drawn from a running vehicle engine on a second-by-second basis. We present one model from each class of models in the following.

6.3.1.1 Emission factor model

These are some of the simpler models where precalculated emission factors are used to estimate the total amount of emissions from a transport activity. There are two ways in which such factor models can be used. The first is *top-down* or *fuel-based*, used when the actual amount of energy or fuel consumption (Ψ_c) relevant to the transport activity is known, such as through fuel receipts, readings from fuel gauges or storage tanks, typically expressed through a unit of energy (e.g. kWh) or fuel (e.g. gallons). The emissions factor (Ψ_f) is the rate of emissions produced per unit of energy or fuel needed to perform the transport activity and is measured in terms of amount per unit of energy (e.g. g/kWh or g/kg). The calculation can be done as follows:

$$\Psi = \Psi_c \times \Psi_f, \tag{6.1}$$

where Ψ is the total amount of emissions (e.g. in g or kg). Typical values of emission factors, in grams per kilogram of fuel used, are reported in Kontovas and Psaraftis (2016). According to this source, aviation has the lowest factor for CO_2 that is equal to 2,600. This is followed by railways (using gas oil/diesel), shipping (using marine diesel or gas oil) and heavy goods vehicles used in road transportation (using diesel), all of which have a CO_2 factor that is equal to 3,140. In terms of NOx emissions, aviation once again has the lowest factor that is equal to 8.3, and is followed by heavy goods vehicles in road transportation with a factor equal to 33.37, then by railways with a higher factor equal to 52.4 and then by shipping with even a higher emission factor equal to 78.5.

If real data as to the use of fuel or energy are not available, then one can resort to the alternative means of calculation that is called *bottom-up* or *activity-based*, in which average emission factors are multiplied by the level of activity to estimate the total amount of emissions. One type of average factor used within road transport is kg of CO_2 per vehicle km, which varies depending on the capacity utilisation of the vehicle. Some conversion factors used in the United Kingdom as of 2013 for various types of vehicles assuming average load are reported by Piecyk (2010). This source provides conversion factors in two types; one that is expressed in kilograms of CO_2 per vehicle kilometre, which would have to be multiplied by the total vehicle kilometres travelled by a vehicle to be able to estimate the total CO_2 emissions. The relevant conversion factors in this case are approximately 0.59 for a rigid truck with a GVWR between 3.5 and 7.5 tonnes, 0.73 for a rigid truck with a GVWR between 7.5 and 17 tonnes and 0.97 for a rigid truck with a GVWR higher than 17 tonnes. The second type of conversion factor is expressed in kilograms of CO_2 per tonne-kilometre, and would have to be multiplied by the total tonne-kilometres travelled to estimate the total emissions. The relevant factors for the same type of rigid trucks mentioned above are 0.58, 0.35 and 0.19, respectively, which, contrary to the first type of conversion factor, decrease with increasing vehicle weight.

6.3.1.2 Average speed model

Of the more popular average speed model is the one described in a report that is published by the European Commission in 1999, which describes methodologies for estimating air pollutant emissions from transport. The model described in the report is named MEET, which estimates rate of emissions $\Psi(v)$ (g/km) for an unloaded goods vehicle on a road with a zero gradient as a function of the average speed v (km/h) of the vehicle:

$$\Psi(v) = \epsilon_0 + \epsilon_1 v + \epsilon_2 v^2 + \epsilon_3 v^3 + \epsilon_4/v + \epsilon_5/v^2 + \epsilon_6/v^3, \tag{6.2}$$

where K and $\epsilon_0 - \epsilon_6$ are predefined coefficients for different types of vehicles classified according to their weight. As an example, the rate of CO_2 emissions (g/km) according to this model for a vehicle of less than 3.5 tonnes weight is given by $\Psi(v) = 0.0617v^2 - 7.8227v + 429.51$. For other classes, the suggested parameters as were calibrated in 1999 by the European Commission are shown in Table 6.1. Figure 6.1 shows the rate of emissions as a function of vehicle speed for vehicles of different weight using the parameters of Table 6.1. The curves seen in the figure exhibit the typical shape of such functions, in which there is often an optimal speed (or a range of optimal speeds) where the rate of emissions is minimised, and decreasing or increasing the speed results in higher emissions.

This model, in its initial form (6.2), was put forward by assuming zero road gradient and for empty vehicles. If the road and vehicle conditions are

Table 6.1 Emission parameters used in MEET

Weight (W)	ϵ_0	ϵ_1	ϵ_2	ϵ_3	ϵ_4	ϵ_5	ϵ_6
3.5 t < W ≤ 7.5 t	110	0	0	0.000375	8702	0	0
7.5 t < W ≤ 16 t	871	−16.0	0.143	0	0	32,031	0
16 t < W ≤ 32 t	765	−7.04	0	0.000632	8334	0	0
W > 32 t	1576	−17.6	0	0.00117	0	36,067	0

Source: Reprinted from *European Journal of Operational Research*, 237(3), E. Demir, T. Bektaş, G. Laporte, A review of recent research on green road freight transportation, pages 775–793, 2014, with permission from Elsevier.

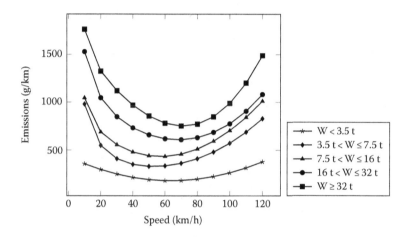

Figure 6.1 Emission rates for trucks of various weights according to the MEET model.

different, a number of corrections may be needed to account for the effects of road gradient and vehicle load on the emissions once a rough estimate has been produced.

Road gradient correction is done by the formula $G(v) = \hat{e}_6 v^6 + \hat{e}_5 v^5 + \hat{e}_4 v^4 + \hat{e}_3 v^3 + \hat{e}_2 v^2 + \hat{e}_1 v + \hat{e}_0$. Similarly, the load factor correction is calculated as follows. $L(\theta, v) = \tilde{e}_0 + \tilde{e}_1 \theta + \tilde{e}_2 \theta^2 + \tilde{e}_3 \theta^3 + \tilde{e}_4 v + \tilde{e}_5 / v^2 + \tilde{e}_6 / v^3 + \tilde{e}_6 / v$, where θ is the value of the road gradient. Some suggested values for the coefficients of both correction factors can be found in Tables 6.2 and 6.3 for CO_2.

Using the correction factors as described earlier, the amount of emissions for a vehicle travelling at speed v and distance τ with a road gradient θ can be calculated as follows:

$$\Psi'(v, D, \theta) = \Psi(v) G(v) L(\theta, v) \tau, \tag{6.3}$$

Table 6.2 Road gradient factors for MEET

Weight (W)	$\hat{\epsilon}_6$	$\hat{\epsilon}_5$	$\hat{\epsilon}_4$	$\hat{\epsilon}_3$	$\hat{\epsilon}_2$	$\hat{\epsilon}_1$	$\hat{\epsilon}_0$	Slope
W ≤ 7.5 t	0	−3.01E−09	5.73E−07	−4.13E−05	1.13E−03	8.13E−03	9.14E−01	[0, 4]
W ≤ 7.5 t	0	−1.39E−10	5.03E−08	−4.18E−06	1.95E−05	3.68E−03	9.69E−01	[−4, 0]
7.5 t < W ≤ 16 t	0	−9.78E−10	−2.01E−09	1.91E−05	−1.63E−03	5.91E−02	7.70E−01	[0, 4]
7.5 t < W ≤ 16 t	0	−6.04E−11	−2.36E−08	7.76E−06	−6.83E−04	1.79E−02	6.12E−01	[−4, 0]
16 t < W ≤ 32 t	0	−5.25E−09	9.93E−07	−6.74E−05	2.06E−03	−1.96E−02	1.45E+00	[0, 4]
16 t < W ≤ 32 t	0	−8.24E−11	2.91E−08	−2.58E−06	5.76E−05	−4.74E−03	8.55E−01	[−4, 0]
W > 32 t	0	−2.04E−09	4.35E−07	−3.69E−05	1.69E−03	−3.16E−02	1.77E+00	[0, 4]
W > 32 t	0	−1.10E−09	2.69E−07	−2.38E−05	9.51E−04	−2.24E−02	9.16E−01	[−4, 0]

Source: Reprinted from European Journal of Operational Research, 237(3), E. Demir, T. Bektaş, G. Laporte, A review of recent research on green road freight transportation, pages 775–793, 2014, with permission from Elsevier.

Table 6.3 Load correction factors for MEET

Weight (W)	$\tilde{\epsilon}_0$	$\tilde{\epsilon}_1$	$\tilde{\epsilon}_2$	$\tilde{\epsilon}_3$	$\tilde{\epsilon}_4$	$\tilde{\epsilon}_5$	$\tilde{\epsilon}_6$	$\tilde{\epsilon}_7$
W ≤ 7.5 t	1.27	0.0614	0	−0.00110	−0.00235	0	0	−1.33
7.5 t < W ≤ 16 t	1.26	0.0790	0	−0.00109	0	0	−2.03E−7	−1.14
16 t < W ≤ 32 t	1.27	0.0882	0	−0.00101	0	0	0	−0.483
W > 32 t	1.43	0.121	0	−0.00125	0	0	0	−0.916

Source: Reprinted from *European Journal of Operational Research*, 237(3), E. Demir, T. Bektaş, G. Laporte, A review of recent research on green road freight transportation, pages 775–793, 2014, with permission from Elsevier.

which, using the values shown in Tables 6.2 and 6.3, returns the total CO_2 in grams.

6.3.1.3 Instantaneous emissions model

Instantaneous models, on the other hand, only deal with 'hot' emissions, that is exhaust emissions of a running engine, and aim to estimate emission rates of an operating vehicle for short time intervals of its driving cycle, for example, on a second-by-second basis. One energy-related emissions estimation model, called the *instantaneous fuel consumption model*, or *instantaneous model* in short, uses vehicle characteristics, such as mass, energy, efficiency parameters, drag force and fuel consumption components associated with aerodynamic drag and rolling resistance, and approximates the fuel consumption per second. The model assumes that changes in acceleration and deceleration levels occur within a one-second time interval. The instantaneous model is shown in the following:

$$\psi_t = \begin{cases} \bar{\Psi} + \beta_1 \Lambda_t v + \beta_2 \bar{\omega} a^2 v / 1000 & \text{for } \Lambda_t > 0 \\ \bar{\Psi} & \text{for } \Lambda_t \leq 0, \end{cases} \tag{6.4}$$

where

ψ_t is the fuel consumption in millilitres per second (mL/s)

$\bar{\Psi}$ is the constant fuel consumption rate of an idle running engine (mL/s)

Λ_t is the total tractive force in kiloNewtons (kN) required to move the vehicle and calculated as the sum of force induced by drag, inertia and road grade

In this function, β_1 is the fuel consumption in millilitres per kiloJoules (mL/kJ) and β_2 is the fuel consumption per unit of energy-acceleration mL/(kJ m/s²). The grade force Λ_t is further calculated as $\Lambda_t = \beta_3 + \beta_4 v^2 + \bar{\omega}\sigma/1{,}000 + \sigma_g \bar{\omega}\theta/100{,}000$, where β_3 is rolling drag force (kN), β_4 is the rolling aerodynamic force (kN/(m/s²)), σ is instantaneous acceleration (m/s^2), $\bar{\omega}$ is the total vehicle weight (kg), v is the speed (m/s), θ is the percent grade and σ_g is the gravitational force (m/s^2). The model operates at a microscale level and is better suited to short trip emission estimations.

A more comprehensive instantaneous emissions model (CMEM) for heavy-duty vehicles (vehicles with a maximum operating mass of 11,794 kg or above) also exists, which models the fuel rate in grams per second (g/s) as follows:

$$\frac{\gamma_r(\gamma_e v_e \gamma_d + \Phi/\epsilon)}{\epsilon_c}, \tag{6.5}$$

where

γ_r is fuel-to-air mass ratio
γ_e is the engine friction factor
v_e is the engine speed
γ_d is the engine displacement
Φ is the second-by-second engine power output in kiloWatts (kW)
ϵ is an efficiency parameter for diesel engines
ϵ_c is a constant

Engine speed v_e is approximated by the vehicle speed v. Once the fuel rate is computed in grams per second, a conversion factor β_c is then applied to the fuel rate to convert it to litres per second (L/s).

The engine power output is calculated as $\Phi = \Phi_t/\epsilon_t + \Phi_a$, where ϵ_t is the vehicle drive train efficiency, and Φ_a is the engine power demand associated with running losses of the engine and the operation of vehicle accessories such as usage of air conditioning. In the rest of the chapter, we will assume $\Phi_a = 0$ for the sake of simplicity. The total tractive power requirement Φ_t placed on the vehicle at the wheels (kW) is calculated as follows:

$$\Phi_t = \frac{(\bar{\omega}\sigma + \bar{\omega}\sigma_g \sin\theta + 0.5e_d\gamma_a\iota v^2 + \bar{\omega}\sigma_g e_r \cos\theta)v}{1000}. \tag{6.6}$$

As seen from (6.6), Φ_t depends on a variety of parameters, including air density γ_a (kg/m^3), frontal surface area of the vehicle ι (m^2), coefficients of aerodynamic drag e_d and rolling resistance e_r, in addition to those already defined earlier. In this equation, the total vehicle weight $\bar{\omega}$ is the sum of the curb (empty) weight ω_c of the vehicle and any load carried on it.

The parameters used in CMEM can be classified as (1) vehicle-independent and (2) vehicle-dependent, typical values for which are given in Tables 6.4 and 6.5, respectively. In particular, the data shown in Table 6.5 are for three different types of vehicles, namely two types of light-duty (LD1 and LD2) and one type of medium-duty (MD) trucks.

Table 6.4 Typical parameters used in the instantaneous model (6.5)

Notation	Description	Typical values
γ_r	Fuel-to-air mass ratio	1
σ_g	Gravitational constant (m/s^2)	9.81
γ_d	Air density (kg/m^3)	1.2041
e_r	Coefficient of rolling resistance	0.01
ϵ	Efficiency parameter for diesel engines	0.45
ϵ_t	Vehicle drive train efficiency	0.45
ϵ_c	Heating value of a typical diesel fuel (kJ/g)	44
β_c	Conversion factor (g/s to L/s)	737

Source: Reprinted from *Transportation Research Part B: Methodological*, 70, Ç. Koç, T. Bektaş, O. Jabali, G. Laporte, The fleet size and mix pollution-routing problem, pages 239–254, 2014, with permission from Elsevier.

Table 6.5 Vehicle-specific parameters

Notation	Description	LD1	LD2	MD
ω_c	Curb weight (kg)	3500	4500	5500
π	Maximum payload (kg)	4000	7500	12500
γ_e	Engine friction factor (kJ/rev/L)	0.25	0.23	0.20
v_e	Engine speed (rev/s)	38.34	37.45	36.67
γ_d	Engine displacement (L)	4.5	4.5	6.9
e_d	Coefficient of aerodynamics drag	0.6	0.64	0.7
ι	Frontal surface area (m^2)	7.0	7.4	8.0

Source: Reprinted from *Transportation Research Part B: Methodological*, 70, Ç. Koç, T. Bektaş, O. Jabali, G. Laporte, The fleet size and mix pollution-routing problem, pages 239–254, 2014, with permission from Elsevier.

6.4 Green freight distribution planning

One way in which externalities can be accounted for in freight distribution and logistics planning is to explicitly incorporate the amount of an externality within the existing models. However, this is not so easy a task for externalities that are difficult to quantify, such as toxic effects or vibration. In this case, it may be possible to factor in estimated costs of such externalities, but the success of this approach depends on the quality and the availability of the estimated costs.

The task is less difficult for other types of externalities for which quantitative models exist that allow the calculation or estimation of the quantity of the externality concerned, and for which cost estimations exist. Greenhouse gas emissions are a good example for such types of externalities, where some of the models introduced in Section 6.3 can be used as estimators of the amount of gases emitted, and for which various cost estimations have been proposed. In this case, the quantitative models can be integrated within the optimisation models, where some of the input parameters that are controllable by a transport planner can be described as decision variables within the optimisation model. In this section, we present two types of problems that take such an approach.

6.4.1 Pollution-routing problems

The first type of problem presented here arises mainly in routing vehicles in road freight transportation and aims to minimise both internal (operational) and external costs, namely those of fuel consumption, the resulting emissions, and driver wages. It is called the pollution-routing problem (PRP), which has been introduced as an extension of the vehicle-routing problem with time windows, and uses the CMEM to estimate the fuel consumption and emissions within vehicle routing. The attraction of CMEM stems from the fact that it includes several input variables, namely distance, load and speed, as well as those related to the type of vehicles chosen in a fleet, all of which can be part of the decisions made by a transport planner.

The model assumes that the fuel consumption of a vehicle with a curb (empty) weight ω_c and carrying a load f, travelling at a constant speed v on a given arc of length d, can be estimated using the following expression:

$$\Phi \approx \Phi_t d/v \tag{6.7}$$

$$\approx (\sigma + \sigma_g \sin \theta + \sigma_g e_r \cos \theta)(\omega_c + f)d \tag{6.8}$$

$$+ (0.5 e_d \iota \gamma_a) v^2 d. \tag{6.9}$$

The expression (6.7) is divided into two terms: Equation 6.8, which shows the part of emissions induced primarily by total vehicle weight $\bar{\omega} = \omega_c + f$, and (6.9), which shows the other part that is fundamentally induced by speed v. For a given fleet of vehicles, note that vehicle-specific parameters such as e_d and ι and external parameters such as σ_g and e_r are fixed and uncontrollable. However, other parameters such as v and f can be controlled, which is the basic principle behind the PRP and using which fuel consumption and emissions can be reduced. To simplify the exposition, we use $\alpha'_{ij} = \sigma + \sigma_g \sin \theta + \sigma_g e_r \cos \theta$ is a vehicle-arc-specific constant and $\hat{\beta} = 0.5 e_d \gamma_a \iota$ is a vehicle-specific constant, where all the input parameters in the calculation of α'_{ij} relate to the arc (i, j) on which a vehicle is traversing.

The PRP is defined on a complete directed graph $G = (V, A)$, where $V = \{0, \ldots, n\}$ is the set of nodes, 0 is the depot and $A = \{(i, j) : i, j \in A \text{ and } i \neq j\}$ is the set of arcs. The distance from i to j is denoted by d_{ij}. A fixed-size fleet of homogeneous vehicles denoted by the set K is available, and each vehicle has capacity π. The set $V_C = V \backslash \{0\}$ is the customer set, and each customer $i \in V_C$ has a nonnegative demand q_i as well as a time interval $[a_i, b_i]$ for service.

Early arrivals at the nodes are permitted but the vehicle has to wait until time a_i before service can start. The service time of customer i is denoted by s_i. The formulation works with a discretised speed function defined by a set $R = \{1, \ldots, |R|\}$ non-decreasing speed levels. Each $r \in R$ corresponds to a fixed average speed $\bar{v}^r$, with $\bar{v}^1$ corresponding to the minimum and $\bar{v}^{|R|}$ to the maximum speeds that a vehicle can use, as allowed by traffic rules.

The PRP aims to find a set of routes for all the vehicles, along with the optimal speeds to traverse each arc of each route, to minimise an objective function that includes the costs of fuel consumption and emissions as calculated by (6.5) and (6.7), as well as the driver costs as a function of time. The parameters c_f and c_d are used to denote the per litre cost of fuel and hourly driver wage, respectively.

The PRP can be formulated using a set of binary and continuous variables. The first set includes variables x_{ij}, each of which is equal to 1 if and only if arc (i, j) appears in solution, and 0 otherwise, and variables z_{ij}^r, each of which is equal to 1 if arc $(i, j) \in A$ is traversed at a speed level r, and 0 otherwise. The second set includes variables f_{ij}, each of which represents the total amount of flow on each arc $(i, j) \in A$, variables y_j, each of which represents the time at which service starts at node $j \in V_C$ and finally variables ρ_j, each of which represents the total time spent on a route that has a node $j \in V_C$ as last visited before returning to the depot. An integer linear programming formulation of the PRP is shown here:

$$\text{Minimise} \quad \sum_{(i,j) \in A} (c_f \gamma_e v_e \gamma_d \hat{\gamma} d_{ij}) \sum_{r \in R} z_{ij}^r / \bar{v}^r \tag{6.10}$$

$$+ \sum_{(i,j) \in A} (c_f \hat{\epsilon} \hat{\gamma} \alpha_{ij}' d_{ij} \omega_c) x_{ij} \tag{6.11}$$

$$+ \sum_{(i,j) \in A} (c_f \hat{\epsilon} \hat{\gamma} \alpha_{ij}' d_{ij}) f_{ij} \tag{6.12}$$

$$+ \sum_{(i,j) \in A} (c_f \hat{\beta} \hat{\epsilon} \hat{\gamma} d_{ij}) \sum_{r=1}^{R} z_{ij}^r (\bar{v}^r)^2 \tag{6.13}$$

$$+ \sum_{j \in V_C} c_d \rho_j \tag{6.14}$$

subject to

$$\sum_{j \in V} x_{0j} = |K| \tag{6.15}$$

$$\sum_{j \in V} x_{ij} = 1 \qquad \forall i \in V_C \tag{6.16}$$

$$\sum_{i \in V} x_{ij} = 1 \qquad \forall j \in V_C \qquad (6.17)$$

$$\sum_{j \in V} f_{ji} - \sum_{j \in V} f_{ij} = q_i \qquad \forall i \in V_C \qquad (6.18)$$

$$q_j x_{ij} \leq f_{ij} \leq (\pi - q_i) x_{ij} \qquad \forall (i,j) \in A \qquad (6.19)$$

$$y_i - y_j + s_i + \sum_{r \in R} d_{ij} z_{ij}^r / \bar{v}^r \leq M_{ij}(1 - x_{ij}) \qquad \forall i \in V, j \in V_C, i \neq j \qquad (6.20)$$

$$a_i \leq y_i \leq b_i \qquad \forall i \in V_C \qquad (6.21)$$

$$y_j + s_j - \rho_j + \sum_{r \in R} d_{j0} z_{j0}^r / \bar{v}^r \leq M(1 - x_{j0}) \qquad \forall j \in V_C \qquad (6.22)$$

$$\sum_{r \in R} z_{ij}^r = x_{ij} \qquad \forall (i,j) \in A \qquad (6.23)$$

$$x_{ij} \in \{0, 1\} \qquad \forall (i,j) \in A \qquad (6.24)$$

$$f_{ij} \geq 0 \qquad \forall (i,j) \in A \qquad (6.25)$$

$$y_i \geq 0 \qquad \forall i \in V_C \qquad (6.26)$$

$$z_{ij}^r \in \{0, 1\} \qquad \forall (i,j) \in A, r \in R. \qquad (6.27)$$

The objective function (6.10) through (6.13) is derived from (6.5) and (6.7), where $\hat{\gamma} = \gamma_r / \epsilon_c \beta_c$ and $\hat{\epsilon} = 1/1000 \epsilon_t \epsilon$ are constants. The terms (6.11) and (6.12) calculate the cost incurred by the vehicle curb weight and payload, respectively. Finally, the term (6.14) measures total driver wages. Constraints (6.15) state that each vehicle must leave the depot. Constraints (6.16) and (6.17) are the degree constraints which ensure that each customer is visited exactly once. Constraints (6.18) and (6.19) define the arc flows. Constraints (6.20) through (6.22) enforce the time window restrictions, where $M_{ij} = \max\{0, b_i + s_i + d_{ij}/\bar{v}^1 - a_j\}$ and M is a sufficiently large number. Constraints (6.23) ensure that only one speed level $r \in R$ is selected for each arc and $z_{ij}^r = 1$ if $x_{ij} = 1$.

6.4.1.1 Fleet mix, fuel consumption and emissions

The PRP presented in Section 6.4.1 assumes a homogeneous fleet of vehicles, each with the same technical parameters. However, different types of vehicles have different rates of fuel consumption and emissions, as was shown in Figure 6.1 for the MEET model, which is also the case for other fuel consumption models discussed in Section 6.3. Therefore, the selection of the type and number of vehicles to be used in a fleet will have an impact on the economic and environmental performance of the distribution plans.

Such considerations give rise to an extension of the PRP in which the aim is to find a set of routes for a heterogeneous fleet of vehicles that meet the demands of all customers within their respective predefined time windows. Each customer is visited once by a single vehicle, each vehicle must depart from and return to the depot, to serve a quantity of demand that does not exceed its capacity. As in the PRP, the speed of each vehicle on each arc must be determined. This problem is called the fleet size and mix pollution-routing problem (FSMPRP), whose objective is to minimise a total cost function encompassing vehicle, driver, fuel and emissions costs.

The FSMPRP is defined on the same graph $G = (V, A)$ as the PRP and uses the same set of input parameters. The only difference is that the index set of vehicle types is denoted by H. Each vehicle of type $h \in H$ has a fixed cost shown by c_h and capacity π^h.

A formulation of the FSMPRP uses a binary variable x_{ij}^h that is equal to 1 if and only if a vehicle of type $h \in H$ travels on arc $(i, j) \in A$. Similar to the PRP, the formulation works with discretised speed levels defined in the set R. The binary variable z_{ij}^{rh} is equal to 1 if and only if a vehicle of type $h \in H$ travels on arc $(i, j) \in A$ at a speed level corresponding to $r \in R$. The continuous variable y_j is the service start time at $j \in V_C$. The total time spent on a route in which $j \in N_0$ is the last visited node before returning to the depot is defined by ρ_j. Furthermore, let f_{ij}^h be the amount of commodity flowing on arc $(i, j) \in A$ by a vehicle of type h. Therefore, the total load of vehicle of type h on arc (i, j) is $\omega_c^h + f_{ij}^h$.

We now present an integer linear programming formulation for the FSMPRP:

$$\text{Minimise} \sum_{h \in H} \sum_{(i,j) \in A} c_f \gamma_e^h v_e^h \gamma_d^h \hat{\gamma} d_{ij} \sum_{r \in R} z_{ij}^{rh} / \bar{v}^r \tag{6.28}$$

$$+ \sum_{h \in H} \sum_{(i,j) \in A} c_f \hat{\gamma} \hat{e}^h \alpha_{ij}' d_{ij} (\omega^h x_{ij}^h + f_{ij}^h) \tag{6.29}$$

$$+ \sum_{h \in H} \sum_{(i,j) \in A} c_f \hat{\gamma} \hat{\beta}^h \hat{e}^h d_{ij} \sum_{r \in R} (\bar{v}^r)^2 z_{ij}^{rh} \tag{6.30}$$

$$+ \sum_{j \in V_C} c_d \rho_j + \sum_{h \in H} \sum_{j \in V_C} c_h x_{0j}^h \tag{6.31}$$

subject to

$$\sum_{j \in V_C} x_{0j}^h \leq m_h \quad \forall h \in H \tag{6.32}$$

$$\sum_{h \in H} \sum_{j \in V} x_{ij}^h = 1 \quad \forall i \in V_C \tag{6.33}$$

$$\sum_{h \in H} \sum_{i \in V} x_{ij}^h = 1 \quad \forall j \in V_C \tag{6.34}$$

$$\sum_{h \in H} \sum_{j \in V} f_{ji}^h - \sum_{h \in H} \sum_{j \in V} f_{ij}^h = q_i \quad \forall i \in V_C \tag{6.35}$$

$$q_j x_{ij}^h \le f_{ij}^h \le (\pi^h - q_i) x_{ij}^h \quad \forall (i,j) \in A, \quad \forall h \in H \tag{6.36}$$

$$y_i - y_j + s_i + \sum_{r \in R} d_{ij} z_{ij}^{rh} / \bar{v}^r \le M_{ij}(1 - x_{ij}^h) \quad \forall i \in V, j \in V_C, \tag{6.37}$$

$$i \ne j, \quad \forall h \in H$$

$$a_i \le y_i \le b_i \quad \forall i \in V_C \tag{6.38}$$

$$y_j + s_j - p_j + \sum_{r \in R} d_{j0} z_{j0}^{rh} / \bar{v}^r \le L_j(1 - x_{j0}^h) \quad \forall j \in V_C \tag{6.39}$$

$$\sum_{r \in R} z_{ij}^{rh} = x_{ij}^h \quad \forall (i,j) \in A, \quad \forall h \in H \tag{6.40}$$

$$x_{ij}^h \in \{0, 1\} \quad \forall (i,j) \in A, \quad \forall h \in H \tag{6.41}$$

$$z_{ij}^{rh} \in \{0, 1\} \quad \forall (i,j) \in A, \quad \forall h \in H \tag{6.42}$$

$$r \in R$$

$$f_{ij}^h \ge 0 \quad \forall (i,j) \in A, \quad \forall h \in H \tag{6.43}$$

$$y_i \ge 0 \quad \forall i \in V_C. \tag{6.44}$$

The first three terms of the objective function represent the cost of fuel consumption and of CO_2 emissions. In particular, term (6.28) computes the cost induced by the engine module, term (6.29) reflects the cost induced by the weight module and term (6.30) measures the cost induced by the speed module. Finally, term (6.31) computes the total driver wage and the sum of all vehicle fixed costs. The maximum number of vehicles available for each type is imposed by constraints (6.32). Constraints (6.33) and (6.34) ensure that each customer is visited exactly once. Constraints (6.35) and (6.36) define the flows. Constraints (6.37) through (6.39) are time window constraints, where $M_{ij} = \max\{0, b_i + s_i + d_{ij}/\bar{v}^1 - a_j\}$ and $L_j = \max\{0, b_j + s_j + d_{j0}/\bar{v}^1\}$. Constraints (6.40) impose that only one speed level is selected for each arc. Finally, constraints (6.41) through (6.44) enforce the integrality and nonnegativity restrictions on the variables.

The following example illustrates the difference that a heterogeneous fleet might have on total cost, time and fuel consumption, in comparison to one of a homogeneous fleet.

Example 6.1

A set of 10 locations are to be visited for deliveries by a fleet of vehicles, each of which will start its tour from a depot and return back to the same depot. The locations, amount of deliveries to be made, time windows (in hours) and service times (in hours) are shown in Table 6.6. The distance data between all locations and the depot are shown in Table 6.7. Deliveries can be made by different types of vehicles that can be hired on a daily basis from the market. Two types of LD (LD1 and LD2) and one type of MD vehicles are available for hire, the data for which are shown in Tables 6.4 and 6.5. The vehicles have different daily costs; in particular, the costs for LD1 and LD2 are £42/day and £49/day, whereas that of MD is £60/day. The drivers are paid £7.92/hour of work they perform, which includes driving time and time spent in making the deliveries. The average combined cost of fuel and CO_2 emissions is estimated at £1.4/L. It is assumed, for reasons of simplification, that all roads are reasonably flat (i.e. $\theta = 0$) and that the effects of acceleration is negligible ($\sigma = 0$).

If there is no flexibility in the choice of the types of vehicles, then a solution using a homogeneous fleet uses three LD vehicles of LD1 which perform the (0, 3, 9, 6, 7, 4, 2, 5, 0), (0, 10, 8, 0) and (0, 1, 0), also shown in Figure 6.2. If, however, there is flexibility in choosing different types of vehicles, then a solution with a heterogeneous fleet would only require two tours, namely (0, 5, 2, 4, 7, 6, 9, 3, 10, 8, 0) using MD and (0, 1, 0) using LD1, visually depicted in Figure 6.3. A comparison of the two solutions in terms of total distance, time, fuel consumption and various costs is given in Table 6.8. Detailed schedules of the two solutions are given in Tables 6.9 and 6.10, which show the arrival time at each node and the average speed used in travelling from one node to another.

As the solutions indicate, using a heterogeneous fleet of vehicles affords greater flexibility in capacity usage and consolidation, which helps to reduce the total distance and en-route time, by about 19% and

Table 6.6 Demand and time data used in Example 6.1

Index i	Name	Demand (kg) q_i	Time a_i	Windows (in hours) b_i	Service time (in hours) s_i
0	Kingston-upon-Hull	0	0.00	9.00	0
1	Pocklington	721	0.60	6.15	24
2	Brough	814	0.18	5.85	27
3	Selby	620	0.29	5.67	21
4	Boughton	311	1.42	6.73	10
5	Barton-upon-Humber	167	0.65	6.03	6
6	Darfield	513	1.02	6.70	17
7	Bentley	568	1.22	6.96	19
8	Watton	763	0.97	6.76	25
9	Cudworth	558	1.04	6.68	19
10	Haxby	636	1.47	7.28	21

Table 6.7 Distance data (in km) used in Example 6.1

	0	1	2	3	4	5	6	7	8	9	10
0	0	41.15	25.68	54.2	95.38	15.91	88.96	74.12	26.01	88.181	66.07
1	40.66	0	51.98	32.8	99.87	42.21	75.66	63.88	24.35	72.07	26.25
2	25.01	51.78	0	61.52	74.05	12.89	69.27	52.59	42.91	73.4	76.7
3	54.27	32.75	61.56	0	77.03	51.93	42.93	31.92	49.48	39.5	29.5
4	94.93	100.03	74.07	76.93	0	81.26	55.6	46.1	111.96	61.7	106.35
5	15.83	42.6	12.88	52.34	81.05	0	78.04	61.32	33.73	82.13	67.52
6	88.751	75.7	69.3	43.03	55.21	78.04	0	17.2	90.55	6.52	68.8
7	73.34	63.44	52.48	31.83	46.43	61.22	17.13	0	75.52	21.26	61.25
8	25.99	24.35	43.78	49.53	111.73	34.01	90.55	75.74	0	88.96	48.92
9	88.411	71.74	73.42	39.43	61.39	82.16	6.55	21.32	88.83	0	64.01
10	65.44	26.25	76.76	29.33	106.07	66.99	68.76	61.18	48.92	64.08	0

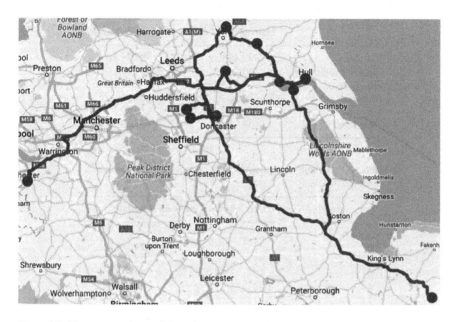

Figure 6.2 The routes obtained for a homogeneous fleet of vehicles. (From Google Maps, 2016.)

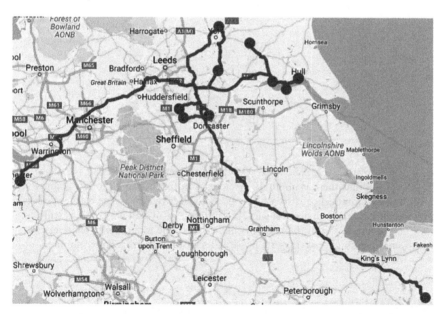

Figure 6.3 The routes obtained for a heterogeneous fleet of vehicles. (From Google Maps, 2016.)

Table 6.8 Comparison of two solutions for Example 6.1

	Homogeneous fleet	Heterogeneous fleet
Distance	489.46 km	398.24 km
Fuel consumption	40.25 L	44.99 L
En-route time	10.32 hours	8.95 hours
Fuel cost	£56.35	£62.99
Driver cost	£81.73	£70.89
Total cost	£138.08	£133.88

Table 6.9 Detailed schedule for the homogeneous fleet used in Example 6.1

Node visited	Arrival time (in hours)	Speed on departure (km/h)
Route 1		
0		75
3	0.72	77
9	1.59	75
6	1.99	75
7	2.51	77
4	3.44	75
2	4.6	74
5	5.22	70
0	6.24	
Route 2		
0		75
10	0.88	76
8	1.89	75
0	2.31	
Route 3		
0		75
1	0.55	75
0	1.77	

13%, respectively. The latter also brings down driver costs. It is noteworthy that there is about 12% increase in fuel consumption in going from a homogeneous to a heterogeneous fleet, which is due to the use of larger (and heavier) vehicles, but in terms of overall cost there is still a reduction. Given the scale of the instance, the reduction is relatively small, but higher savings are possible for instances with a larger number of customers.

6.4.2 Speed optimisation on fixed routes

The previous section described a way in which fuel consumption and greenhouse gas emissions can be factored into route planning, including selection

Table 6.10 Detailed schedule for the heterogeneous fleet used in Example 6.1

Node visited	Arrival time (in hours)	Speed on departure (km/h)
Route 1		
0	0	75
5	0.21	75
2	0.48	76
4	1.92	75
7	2.7	77
6	3.25	75
9	3.62	76
3	4.45	75
10	5.19	74
0	7.18	
Route 2		
0	0	75
1	0.55	75
0	1.77	

of arcs and the speeds with which they are traversed. A related problem arises when the route is fixed, but when the speeds used along the route need to be optimised to reduce fuel consumption. Such a problem arises in maritime transportation, and in particular liner shipping, which operates on the basis of published timetables (or schedules) which indicates that the route they operate on is fixed. These ships therefore follow a given sequence of ports to visit, under the constraints that each port must be visited such that service starts within pre-specified time windows. In this case, the only possibility to reduce fuel consumption lies in changing speeds between each pair of successive ports on the journey, and the problem is to find the optimum speeds on each leg of the journey. The problem is not restricted to the maritime domain and can also be applied to road transport where vehicle routes are fixed, including long-haul transportation, or where repeated deliveries are made to known customer locations over a long period of time, in which case routes are predefined and fixed for the duration of the planning period.

The problem of optimising speeds on a given sequence $1, 2, \ldots, n$ of customers is known as the speed optimisation problem (SOP). Each customer k requires service to start within the time window $[a_k, b_k]$. Each arc $(k, k+1)$ has length $d_{k,k+1}$ and the cost of fuel consumption expressed as a function $F(v_{k,k+1})$ of speed $v_{k,k+1}$ chosen on that arc, where $k = 1, 2, \ldots, n-1$. We assume that the individual arc costs are additive. The objective of SOP is to minimise a total cost function that is a summation of the costs of all arcs, and possibly other components (such as driver or crew costs).

The travel time on an arc $(k, k+1)$ as $t_{k,k+1} = d_{k,k+1}/v_{k,k+1}$ under the assumption that speed $v_{k,k+1}$ is kept constant on the arc $(k, k+1)$. This may not be such a strong assumption for maritime shipping where speed is not subject to significant changes unless there are unforeseen conditions such as weather. However, it would be a strong assumption for road transport where vehicle speeds are subject to regular changes due to traffic conditions. In this case, $v_{k,k+1}$ can be interpreted as the *average* speed on arc $(k, k + 1)$. Speeds are also restricted to be between a minimum v_{min} and a maximum v_{max} value.

With these definitions, a formulation for the SOP is presented here:

$$\text{Minimise} \quad \sum_{k=1,\ldots,n-1} d_{k,k+1} F(v_{k,k+1}) \tag{6.45}$$

subject to

$$t_{k+1} - t_k - d_{k,k+1}/v_{k,k+1} \geq 0 \qquad \forall k = 1,\ldots,n-1 \tag{6.46}$$

$$a_k \leq t_k \leq b_k \qquad \forall k = 1,\ldots,n \tag{6.47}$$

$$v_{min} \leq v_{k,k+1} \leq v_{max} \quad \forall k = 1,\ldots,n-1. \tag{6.48}$$

The objective function (6.45) minimises the total fuel consumption on the given route, where $f(v)$ is the fuel consumption as a function of speed v. Constraints (6.46) are used to restrict the start time of service at a given node only after the ship has arrived to that node. Time window restrictions are modelled through constraints (6.47), whereas the last set of constraints (6.48) enforce minimum and maximum speed limits on each arc as v_{min} and v_{max}, respectively. $F(v)$ is, in general, a quadratic convex function, which makes the model non-linear in the objective function and in the constraints. Fagerholt et al. (2010) show that, by using travelling time as a decision variable, the non-linearity in the constraints can be avoided. Furthermore, by discretising the arrival times at each node and replicating each node for each possible arrival time, much in the same way as a time–space network configuration. In this way, the SOP can be modelled and solved as the shortest path problem on a directed acyclic graph.

In the following section, we show how this formulation can be applied to a speed optimisation problem arising in maritime transportation in the following case study drawn from Fagerholt et al. (2010).

6.5 Practical application: Speed optimisation in maritime shipping

This instance is described by Fagerholt et al. (2010), in which a ship is scheduled to sail on a route with the seven ports labelled as $k = 1,\ldots,7$, and correspond to the following locations, Antwerp, Milford Haven, Boston, Charleston, Algeciras, Point Lisas and Houston, respectively. The distance, in nautical miles, between two subsequent ports $(k, k+1)$ on the given route is shown by $d_{k,k+1}$. The distances are as follows: $d_{1,2} = 510$, $d_{2,3} = 2699$, $d_{3,4} = 838$, $d_{4,5} = 3625$, $d_{5,6} = 3437$ and $d_{6,7} = 2263$. The time windows within which the ship must arrive at each port $k = 1, \ldots, 7$, indicated by the earliest a_k and latest b_k times, are shown as $[a_k, b_k]$. For the first port $k = 1$, the time window is set as $[a_1, b_1] = [0, 0]$, which indicates that the ship is ready to sail and therefore starts her journey immediately. For this instance, the time windows for the remaining ports, in days, are given as follows: $[a_2, b_2] = [1, 5]$, $[a_3, b_3] = [9, 13]$, $[a_4, b_4] = [11, 15]$, $[a_5, b_5] = [20, 24]$, $[a_6, b_6] = [32, 36]$ and $[a_7, b_7] = [35, 39]$. According to the constraints posed by the time windows, for example, the ship cannot access port 2 before day 1 and has to reach the same port by day 5 at the latest. If the ship arrives at a port k earlier than day a_k, then it has to wait idle until day a_k, and only then access to the port will be possible. Similarly, any arrival later than day b_k would not be acceptable, and any schedule with such late arrivals would be considered infeasible. For the purposes of this illustrative example, it is assumed that the time spent at each port is negligible, although in practice dwell times will have an impact on the schedule.

Fagerholt et al. (2010) derive the following fuel consumption function using real data from a particular ship:

$$F(v) = 0.0036v^2 - 0.1015v + 0.8848,$$

where

  v is the speed of the ship (in knots, where one knot is equal to one nautical mile per hour)

  $F(v)$ is the rate of consumption in tonnes per nautical mile

The speed of the ship can be changed from one segment of the route to another but will have to remain constant on each leg. The minimum and maximum speeds that the ship can sail at are 14 and 20 knots, respectively.

If the ship sails at a constant speed throughout its voyage, the total fuel consumption for different values of speed would be as follows:

$$\sum_{k=1,\ldots,6} d_{k,k+1} F(18) = 2998 \text{ tonnes}$$

$$\sum_{k=1,\ldots,6} d_{k,k+1} F(18.5) = 3197.91 \text{ tonnes}$$

$$\sum_{k=1,\ldots,6} d_{k,k+1} F(19) = 3421.89 \text{ tonnes}.$$

A formulation of the speed optimisation problem for this instance is as follows:

Minimise $510(0.0036v_{12}^2 - 0.1015v_{12} + 0.8848)$

$\qquad +2699(0.0036v_{23}^2 - 0.1015v_{23} + 0.8848)$

$\qquad +838(0.0036v_{34}^2 - 0.1015v_{34} + 0.8848)$

$\qquad +3625(0.0036v_{45}^2 - 0.1015v_{45} + 0.8848)$

$\qquad +3437(0.0036v_{56}^2 - 0.1015v_{56} + 0.8848)$

$\qquad +2263(0.0036v_{67}^2 - 0.1015v_{67} + 0.8848)$

subject to

$$t_2 - t_1 - 510/v_{12} \geq 0$$
$$t_3 - t_2 - 2699/v_{23} \geq 0$$
$$t_4 - t_3 - 838/v_{34} \geq 0$$
$$t_5 - t_4 - 3625/v_{45} \geq 0$$
$$t_6 - t_5 - 3437/v_{56} \geq 0$$
$$t_7 - t_6 - 2263/v_{67} \geq 0$$
$$0 \leq t_1 \leq 0$$
$$24 \leq t_2 \leq 120$$
$$216 \leq t_3 \leq 312$$
$$264 \leq t_4 \leq 360$$
$$480 \leq t_5 \leq 576$$
$$768 \leq t_6 \leq 864$$
$$840 \leq t_7 \leq 936$$
$$14 \leq v_{12} \leq 20$$
$$14 \leq v_{23} \leq 20$$
$$14 \leq v_{34} \leq 20$$
$$14 \leq v_{45} \leq 20$$
$$14 \leq v_{56} \leq 20$$
$$14 \leq v_{67} \leq 20.$$

The solution of this formulation is where the ship sails at a speed equal to 14.28 on all legs of the journey, which will result in the following arrival times at ports: $t_2^* = 1.49$ days, $t_3^* = 9.36$ days, $t_4^* = 11.80$ days, $t_5^* = 22.38$ days, $t_6^* = 32.4$ days and $t_7^* = 39$ days, assuming the journey starts on day $t_1^* = 0$, yielding a total fuel consumption figure of around 2266 tonnes.

References and further reading

Green transportation and logistics is a relatively new field of investigation, yet books on the topic already exist given the importance of the research area. An excellent introduction to the topic can be found in the book edited by McKinnon et al. (2015), which covers aspects such as assessing, evaluating and internalising the environmental effects of logistics, strategic and operational issues, as well as related topics such as sustainability for city logistics and reverse logistics. The book edited by Psaraftis (2016) presents a comprehensive coverage of the relevant environmental issues with a particular focus on maritime transportation, but also includes chapters on road and rail transportation, aviation and inland navigation. Demir et al. (2014b) and Eglese and Bektaş (2014) present surveys on green road transportation and green vehicle routing, where the former reference also reviews a number of fuel consumption models. A numerical comparison of six fuel consumption models using simulations can be found in Demir et al. (2011).

The PRP (Bektaş and Laporte, 2011) and its variations have been studied in sufficient detail, see Demir et al. (2012) and Kramer et al. (2015b) for heuristic algorithms, Demir et al. (2014a) for a bi-objective variant of the problem where driver time and fuel consumption have been treated as different and conflicting objectives and Koç et al. (2014) for the problem involving a heterogeneous fleet where the additional decisions entail deciding on the fleet composition for distribution. Vehicle routing with environmental concerns within urban zones has also been subject of some research, in particular Ehmke et al. (2016) study a type of a PRP where emissions are time dependent. Koç et al. (2016a) investigate a broader set of decisions concerning depot location, fleet composition and vehicle routing within an urban setting, where speed restrictions are characterised by speed zones within city centres.

Speed optimisation has been studied earlier within maritime logistics, although not necessarily with environmental concerns in mind. Interested readers may consult the papers by Psaraftis and Kontovas (2013) and Psaraftis and Kontovas (2014). Speed optimisation as a stand-alone problem has been looked at in Fagerholt (2001), and later by Fagerholt et al. (2010) in the context of ship routing and scheduling, for the solution of which Hvattum et al. (2013) present an algorithm. These ideas have later been adapted to the particular characteristics of road transportation by Demir et al. (2012), and later by Franceschetti et al. (2013) and Kramer et al. (2015a) who consider additional decisions relevant to the departure times of the vehicles.

The models presented in this chapter are based on those described in the references above.

Chapter 7

Collaboration in freight distribution

There exist two main types of collaborations within logistics and distribution. The first is *vertical*, which typically arises in supply chains and entails collaboration between different levels of the chain, typically involving actors that have distinct and nonoverlapping roles within the chain. For example, transport managers may collaborate with warehouse managers to jointly optimise their plans to achieve a better integration of the operations, resulting in less cost and improved system performance. This particular type of collaboration has been the subject of extensive research within the context of supply chains.

The second type of collaboration is *horizontal*, where providers of the same (or similar) service share resources, such as the network infrastructure (including depots and links), and jointly plan the routes and schedules for the services they offer. The goal is to achieve better coordination of the assets, so as to gain a higher efficiency without detracting from the service quality.

Both vertical and horizontal collaborations are relevant to freight logistics, although the latter is more so given that the actors within the collaboration perform fairly similar roles of carrying goods from their origins to destinations. The distinction between the two types depends on the extent to which the roles of the various actors overlap. If, for example, the collaboration between two parcel carriers is such that one exclusively performs deliveries between facilities and warehouses, and the other from warehouses to customers, then vertical collaboration will be more relevant to the two carriers than horizontal one. For two other operators who both perform last mile deliveries, however, the collaboration will typically be horizontal.

For any collaboration to work effectively, and for it to sustain itself over a period of time, it is important that the partners of the collaboration get a 'fair' share of the *pain and gain*. A collaboration will typically have a *value*, that can be expressed as the total profit that the collaboration generates, or the total cost it incurs. Simple allocation rules can be used to calculate the share of each partner from the value of the collaboration, which might be in the form of profits or costs. These rules include

- Dividing the value of the collaboration equally among the partners of the collaboration
- Sharing the value of the collaboration in a way that is proportional to the profit or the cost each partner brings

Unfortunately, using such simple rules do not always result in a fair allocation. Cooperative game theory is a way to answer some of the questions around formation of collaborations and to achieve a fair distribution of the benefits. This chapter is a brief introduction to some of the key concepts used in cooperative game theory, with a particular focus on the application of these concepts to collaborative logistics. In the next section, we formalise some of the concepts introduced earlier.

7.1 Cooperative game theory

In cooperative game theory terminology, a number of *players* form a *coalition* so that they can benefit from working together and achieve a more effective use of their resources.

Given a set N of users who are willing to collaborate in a joint venture, cooperative game theory is concerned with two fundamental questions:

1. How does one find the structure of the coalition formed by a subset $N_C \subseteq N$ of the users?
2. How does one calculate a *payoff* vector x that describes how to divide the value $v(N_C)$ of a coalition among the members in a fair way?

The answer to the second question should be such that payoffs reflect the contribution of each member, and that the coalition is *stable*, meaning that there is no incentive for any member to leave the coalition, or no members of the coalition can find a way of collaborating on their own with a better cost allocation. The latter aspect is also relevant to the ability to sustain the collaboration. Following are some of the concepts through which these questions can be answered.

7.1.1 Core

For a given coalition, if there exists a subset of the coalition with a worse payoff than the value of the subset, then there will be an incentive for some players to leave the coalition and form a new one. The concept of *core* has been introduced as a means of finding a payoff that is stable and one for which there would be no incentive to deviate from a coalition. In particular, for any set N of players, all nonnegative payoff vectors $x = (x_1, x_2, \ldots, x_{|N|})$ that satisfy $x(N) = v(N)$ and $\sum_{i \in N_C} x_i = x(N_C) \leq v(N_C)$ for all subsets $N_C \subset N$ form the core.

The definition of the core implies that it may not be possible to form a stable coalition in a cooperative game, in which case the core will be empty. Checking whether the core of a game is nonempty or not can be done through the use the following linear programming formulation:

Minimise ϵ

subject to

$$x(N_C) \leq v(N_C) + \epsilon \quad \forall N_C \subset N$$

$$x(N) = v(N).$$

In particular, the first set of constraints imply that any subset N_C of the set N of players should always have a better payoff, denoted by the value $x(N_C)$ if they were to stay in the collaboration, as opposed to forming a new collaboration with value $v(N_C)$. The second set of constraints guarantee that the value of the collaboration should be distributed among the members.

If the optimal value of the linear programme mentioned earlier is $\epsilon^* = 0$, then the core of the game is nonempty, in which case an optimal solution $x^* = (x_1^*, \ldots, x_{|N|}^*)$ will correspond to the payoff vector. On the other hand, if $\epsilon^* > 0$, then the core of the game is empty. In this case, x^* is the *least-core* of the game and ϵ^* is called the *least-core* value.

The following example shows a collaboration game with a nonempty core.

Example 7.1

Figure 7.1 shows individual distribution centres, both of which belong to one logistics provider, used to perform deliveries to warehouses. One centre is located at node A and the other located at node B. The warehouses are located on the nodes shown by A_1, A_2 and B_1. The distances between various nodes in the network are indicated next to the links shown. The centres are autonomous in their operation and finances and therefore operate independently. Both centres are supplied by a central facility with the goods required for distribution. For the purposes of this example, we assume that the cost of supply is negligible.

The centre at A performs direct shipments to two warehouses, one at node A_1 located 30 km away and the other at node A_2 located 50 km away. The cost of distribution is proportional to the distance travelled, for which reason total distance is used as an estimate for the total cost.

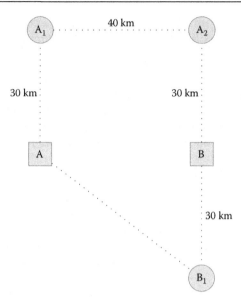

Figure 7.1 The network used in Example 7.1.

The distribution centre at B works only with one warehouse located 30 km away at node B_1. The links used to perform the shipments from the two centres are shown by the thick arcs in Figure 7.2. The total distance for the three sets of shipments is 110 km in this plan.

If the two centres were to collaborate, then a solution shown in Figure 7.3 can be used to perform the distribution operations, where the demand for the warehouse at node A_2 can be met from the centre at node B, reducing the total distance for the shipments down to 90 km.

Let the cost share be denoted by the vector $x = (x_A, x_B)$, where x_A and x_B show the costs allocated to the centres at A and B, respectively. For the collaboration to be stable and fair, the vector x would need to satisfy the following conditions:

$$0 \leq x_A \leq 80, 0 \leq x_B \leq 30, x_A + x_B = 90.$$

It is easy to verify that the core is nonempty, given that points such as $(60, 30)$, $(70, 20)$, $(65, 25)$ satisfy the given set of conditions. It is also easy to see that there is no benefit for either of the distribution centres to leave the coalition and operate individually, provided the costs are allocated in line with the payoff vectors from within the core.

Next, we provide an example of a collaboration where the corresponding game has an empty core.

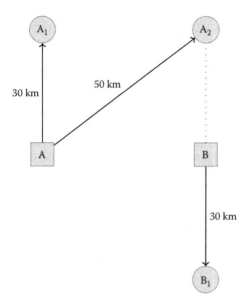

Figure 7.2 Individual distribution plans for the two centres of Example 7.1.

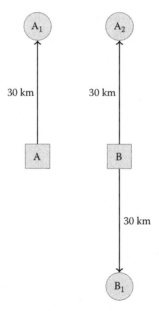

Figure 7.3 Jointly optimised distribution plans assuming collaboration in Example 7.1.

Example 7.2

Three independent parcel carriers regularly deliver consignments to their customers everyday. The carriers are based at depots shown by A, B and C in Figure 7.4. In particular, the carrier based at depot A (or simply player A) has to deliver to customers A_1 and A_2, the carrier based at depot B (player B) has to deliver to customers B_1 and B_2 and the carrier based at depot C (player C) has to deliver to customers C_1 and C_2. Each carrier uses a single van to carry for the deliveries, and the sizes of the consignments are small enough so that the capacity of the van does not pose any limitations or restrictions in performing the deliveries. The cost of travel for the van is proportional to the distance travelled. Each carrier wishes to find a route for their van to depart from and return back to the depot, having visited each of their customers exactly once, such that the total distance is minimised. In other words, each carrier will need to solve a travelling salesman problem to minimise their individual shipment costs.

The figure also shows the possible links that the vans can travel on, each being bidirectional, and indicates the distances for the horizontal and vertical edges. The distances for the diagonal links are calculated using simple geometry. In particular, the shorter diagonal links (e.g. A to C_1) have a value equal to 29 km. The length of each of the links (A, A_1), (C, B_1), (A, B_2) and (C, C_2) is equal to 46.51 km, whereas the length of each of the links (B_1, A_2), (C_1, C_2), (C_1, B_2) and (A_1, A_2) is equal to 45.17 km.

A solution where each carrier optimises their own route is shown in Figure 7.5. In this case, the route for carriers A and C will each be 120.68 km long, and the route length for carrier B is 116 km long, resulting in a total distance of 357.36 km across the three routes.

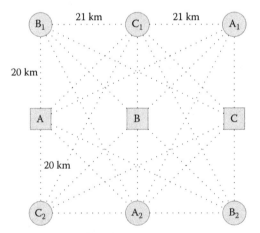

Figure 7.4 The distribution network for Example 7.2.

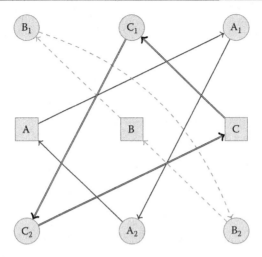

Figure 7.5 Individually optimised solutions for each of the three carriers in Example 7.2.

On the other hand, a solution where the three carriers collaborate to operate by sharing their vehicles and the depots in serving the six customers, such as the one shown in Figure 7.6, will reduce the overall total distance. This is an instance of a multi-depot vehicle routing problem with three depots, three vehicles and six customers, which can be solved using similar formulations to those shown in Chapter 4. The solution shown has a value equal to 240 km in total, which is approximately 33% less than the total distance obtained using the individually optimised solutions. To determine if the core is nonempty, one must first

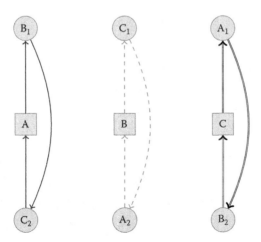

Figure 7.6 A solution for Example 7.2 with all three carriers collaborating.

find the values of all subsets $N_C \subset N = \{A, B, C\}$ and their values, as indicated here:

- $v(\{A\}) = 120.68$ km
- $v(\{B\}) = 116$ km
- $v(\{C\}) = 120.68$ km
- $v(\{A, B\}) = 150$ km (see Figure 7.7)
- $v(\{A, C\}) = 140$ km (see Figure 7.8)
- $v(\{B, C\}) = 150$ km,

using which one can construct the following linear programme:

Minimise ϵ

subject to

$$x_A \leq 120.68 + \epsilon$$

$$x_B \leq 116 + \epsilon$$

$$x_C \leq 120.68 + \epsilon$$

$$x_A + x_B \leq 150 + \epsilon$$

$$x_A + x_C \leq 140 + \epsilon$$

$$x_B + x_C \leq 150 + \epsilon$$

$$x_A + x_B + x_C = 240,$$

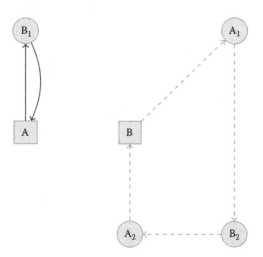

Figure 7.7 A solution for Example 7.2 with carriers A and B collaborating.

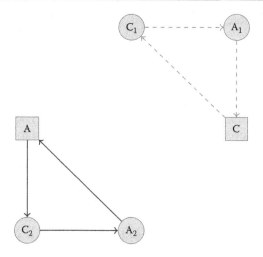

Figure 7.8 A solution for Example 7.2 with carriers A and C collaborating.

which yields an optimal value $\epsilon^* = 13.33$, and a corresponding optimal solution $(x_A^*, x_B^*, x_C^*) = (76.67, 86.67, 76.67)$, indicating that the core is empty and that there is an incentive for a subset of the carriers to deviate from a coalition involving all three carriers. For example, whereas a collaboration $N = \{A, B, C\}$ would leave $\{A, B\}$ with a total share equal to 163.34 km, a collaboration between the two players alone could achieve a lower total distance equal to 150 km. A similar conclusion holds for the subsets $\{A, C\}$ and $\{B, C\}$.

For games with an empty core, there exist alternative mechanisms to use for allocating costs. Two of these are presented here.

7.1.2 Shapley value

For a given coalition N_C, the Shapley value calculates the payoff of each member of the coalition to be proportional to the marginal contribution of that member. If $v(N_C)$ is the value of a subset N_C of the set N of players, the Shapley value of player i is calculated as

$$x_i^S = \sum_{N_C \subseteq N \setminus \{i\}} \frac{|N_C|!(|N| - |N_C| - 1)!}{|N|!} (v(N_C \cup \{i\}) - v(N_C)), \qquad (7.1)$$

where the value $v(N_C \cup \{i\}) - v(N_C)$ is the marginal contribution of player $i \in N$ to the coalition.

The Shapley value has a number of useful properties, some of which are listed as follows:

- *Efficiency*: The value of the coalition is fully distributed among all the players of the game. In other words, if x_i^S is the Shapley value for a player $i \in N$ of a game, then $\sum_{i \in N} x_i^S = v(N)$.
- *No pain, no gain*: The Shapley value of a player who does not contribute to any coalition is 0. Such players are often referred to as *dummy* in the game.
- *Symmetric*: If two players have equal contributions to a coalition, then their Shapley value is the same.

Example 7.3

Let us consider the collaboration in Example 7.2, where it was shown that the core is empty. In this case, one can use the Shapley values to determine a cost allocation. We start by calculating the Shapley value for player A and observe that there are four subsets of N which include this player, namely {A}, {A, B}, {A, C} and {A, B, C}, which correspond to the four following subsets of $N\backslash\{A\}$, namely $N_1 = \emptyset$, $N_2 = \{B\}$, $N_3 = \{C\}$ and $N_4 = \{B, C\}$. Then the Shapley value x_A^S of player A can be calculated as follows:

$$
x_A^S = \frac{|N_1|!(|N| - |N_1| - 1)!}{|N|!} \left(v(\{A\}) - v(\emptyset) \right)
$$

$$
+ \frac{|N_2|!(|N| - |N_2| - 1)!}{|N|!} \left(v(\{A, B\}) - v(\{B\}) \right)
$$

$$
+ \frac{|N_3|!(|N| - |N_3| - 1)!}{|N|!} \left(v(\{A, C\}) - v(\{C\}) \right)
$$

$$
+ \frac{|N_4|!(|N| - |N_4| - 1)!}{|N|!} \left(v(\{A, B, C\}) - v(\{B, C\}) \right),
$$

which translates into

$$
x_A^S = \frac{1}{3} (120.68 - 0)
$$

$$
+ \frac{1}{6} (150 - 116)
$$

$$
+ \frac{1}{6} (140 - 120.68)
$$

$$
+ \frac{1}{3} (240 - 150)
$$

$$
= 79.1133.
$$

Using similar calculations, the Shapley value of players B and C can be computed as $x_B^S = 81.7733$ and $x_C^S = 79.1133$. The latter is a direct consequence of the symmetric nature of players A and C, which implies that $x_A^S = x_C^S$, without needing further calculation. Finally, it can be verified that the efficiency property holds by observing that $x_A^S + x_B^S + x_C^S = 240$.

7.1.3 Banzhaf index

The Banzhaf index is a similar measure to the Shapley value, defined individually with each player, and is calculated as follows:

$$\frac{1}{2^{|N|-1}} \sum_{N_C \subseteq N \setminus \{i\}} (v(N_C \cup \{i\}) - v(N_C)), \tag{7.2}$$

where the total marginal contributions are averaged across all coalitions of the game. One difference between the Banzhaf index and the Shapley value is that the former does not have the efficiency property.

Other concepts such as the nucleolus, the kernel, the bargaining set and the stable set also exist, for which detailed descriptions can be found in, for example Chalkiadakis et al. (2012).

7.2 Practical application: Cooperation among freight carriers

This section is based on the work of Krajewska et al. (2008), describing a freight forwarding company operating across Germany which consists of 19 centres, each of which operates as a freight carrier. The company receives a set of requests on a daily basis, where each request is specified by (1) a pickup node, where a certain commodity is loaded onto a vehicle, (2) a delivery node to which the commodity is transported and (3) the weight of the commodity. Each node in the network, whether pickup or delivery, is associated with a service time and a time window specified by the earliest and latest times by which service can commence at this node. Each centre operates its own fleet of vehicles, with the fleet size ranging from 5 to 15 vehicles, and acts independently. In a non-cooperative setting, the company partitions the requests received and assigns them to the carriers. Each carrier then solves a pickup and delivery vehicle routing problem with time windows in order to optimise the route of their vehicles. A significant issue reported with this approach is that a large portion of the vehicle movements within each centre is empty.

The case study reported by Krajewska et al. (2008) has looked at the possibility of three carriers collaborating and used an instance of the problem with 257 requests and 38 vehicles to investigate the potential benefits of such a collaboration. The collaboration has given rise to a multi-depot pickup and delivery problem with time windows, which is solved using a heuristic algorithm. It is shown that the collaboration, in comparison to a non-cooperative scenario, achieves reductions in cost that range from 9% to 19%. The Shapley value has been used as a means of allocating the total cost between the three players. The resulting payoff vector is shown to be stable, fair and one that is in the core, where one player achieves about a 20% savings in cost and the other two with savings in the region achieve 10% each.

References and further reading

Collaboration in logistics is an area that is still under-researched. The paper by Krajewska and Kopfer (2006) and the book chapter by Agarwal et al. (2009) serve as a good introduction. A review with a focus on road transportation is presented by Verdonck et al. (2013). Models and applications are presented by Özener and Ergun (2008) in full truckload transportation and by Krajewska et al. (2008) and Wang and Kopfer (2014) in less-than-truckload freight transportation. For an introduction to cooperative game theory and computational aspects, we refer the reader to the book by Chalkiadakis et al. (2011). A case study arising in forestry is described in Frisk et al. (2010), including the description of a new allocation mechanism.

The models presented in this chapter are based on those described in the references above.

Chapter 8

Methodology

The focus of this book is on formulating various optimisation problems arising in freight transportation and distribution using mathematical models, and in particular integer linear programming formulations. For this reason, this chapter will introduce a number of ways in which they can be solved. Some of the material covered in this chapter will require basic familiarity with linear programming, for which references are provided at the end of this chapter.

We assume all the material covered in this section pertains to a single-objective mathematical programming formulation, which we show by $\mathcal{F}$ and state as follows in its general form:

$(\mathcal{F})$ Minimise $f(x)$

subject to

$$g(x) \geq b$$
$$x \in \mathbb{R}^n \times \mathbb{Z}^m.$$

In this formulation, $f(x)$ is a continuous function of the vector $x = (x_1, x_2, \ldots, x_{n+m})$ of variables, and b is a vector of right-hand-side coefficients.

We now differentiate between the following types of mathematical models:

- If $f(x)$ and $g(x)$ are both linear, then
  - If $n \geq 1$ and $m = 0$, then $\mathcal{F}$ is called a linear programme (LP).
  - If $n = 0$, $m \geq 1$, then $\mathcal{F}$ is called a (pure) integer programme (IP).
  - If $n \geq 1$ and $m \geq 1$, then $\mathcal{F}$ is called a mixed-integer programme.

- If either $f(x)$ or $g(x)$ is non-linear, then
  - If $n \geq 1$ and $m = 0$, then $\mathcal{F}$ is called a non-linear programme (NLP).
  - If $n = 0$, $m \geq 1$, then $\mathcal{F}$ is called a (pure) integer non-linear programme (INLP).
  - If $n \geq 1$ and $m \geq 1$, then $\mathcal{F}$ is called a mixed-integer non-linear programme (MINLP).
  - If both $f(x)$ and $g(x)$ are convex, then $\mathcal{F}$ is a convex programme.
  - If either $f(x)$ or $g(x)$ is quadratic, then $\mathcal{F}$ is quadratic programme (QP).

Optimisation problems with integer variables are *combinatorial* in nature, meaning that there are a finite number of solutions. Such problems are also referred to as *discrete optimisation* problems, which consist of finding an optimal solution from the finite set of solutions. Practical discrete optimisation problems typically have a prohibitively large number of feasible solutions that make an exhaustive search impractical.

8.1 General-purpose solvers

A relatively straightforward way of solving mathematical programming formulations is to use general purpose solvers, including those that are either commercially available or free for public use. Modern solvers have become very powerful in solving optimisation problems and are likely to provide optimal solutions for models of reasonable size, depending on the type of problem being solved. Even if the models are unable to be solved optimally, the solvers will typically return a feasible solution for the problem at hand, as well as a lower bound for the user to be able to assess the quality of the solution identified.

In the following, we list some of the solvers available for this purpose:

- CLP is a freely available and an open-source solver for linear programmes, including those with linear or quadratic objectives, and is available either as a callable library or a stand-alone application (https://projects.coin-or.org/Clp).
- Solver is a numerical optimisation tool that is available as an add-in within Microsoft Excel, able to solve linear and mixed-integer models using the simplex method and non-linear programming models using genetic algorithms (GAs) and local search (http://www.solver.com by Frontline Systems).
- The IBM CPLEX Optimizer is a commercial mathematical programming solver for linear, mixed-integer and quadratic programming models (https://www.ibm.com/software/commerce/optimization/cplex-optimizer/ by IBM). It is available as a stand-alone interactive

optimiser or can be used through an algebraic modelling language or as callable libraries within applications.

- GUROBI is a commercial mathematical programming solver for linear, mixed-integer and quadratic programming models (http://www.gurobi.com by GUROBI Optimization). It is available as a stand-alone interactive optimiser or can be used through an algebraic modelling language or as callable libraries within applications.
- FICO® Xpress Optimization Suite is a commercial package for solving large-scale linear, mixed-integer and non-linear models, as well as constraint programming problems, which comes with a programming language that allows users to interact with the solver engines (http://www.fico.com/en/products/fico-xpress-optimization-suite).
- BONMIN (http://www.coin-or.org/Bonmin/) is a freely available open-source solver for general non-linear integer programming formulations, possibly with integer variables. It is able to solve general mixed integer non-linear programmes to optimality if they are convex in the objective function and the constraints, otherwise, it returns a heuristic solution. It incorporates algorithms such as a non-linear programming based branch-and-bound, outer approximation and a hybrid branch-and-cut.
- COUENNE (https://projects.coin-or.org/Couenne) is a freely available open-source solver for general non-linear integer programming formulations, possibly with integer variables. In contrast to BONMIN, it is able to solve non-convex mixed integer non-linear programmes, using tools such as linearisation, bound reduction and branching methods within branch-and-bound.
- BARON (Branch-And-Reduce Optimization Navigator) (http://archimedes.cheme.cmu.edu/?q=baron) is a general purpose solver for purely continuous, purely integer and mixed-integer non-linear problems, which uses constraint propagation, interval analysis and duality within branch-and-bound.

Most of the solvers listed and many more are available on the NEOS Server (https://neos-server.org/neos/), which is a free Internet-based service for solving optimisation models. It has an excellent collection of state-of-the-art solvers for different types of models, but also includes problem-specific solvers. The website allows users to input their models using a variety of formats, which is then solved on a distributed cluster of high-performance machines, and a solution is returned to the user via the same web interface. Another excellent repository of open-source software is COIN-OR (http://www.coin-or.org), which includes a wealth of algorithms, ranging from solvers for linear, integer and non-linear programming to libraries of algorithms for particular problems, such as the vehicle routing problem.

One potential limitation of using general purpose solvers to solve optimisation problems is when the models are of large scale. In this case, one can resort to more specialised solution techniques that rely on decomposing or relaxing the formulations, such as the ones explained in the following section.

8.2 Exact solution techniques

Practical applications of linear and integer programming models often give rise to large-scale formulations, mainly due to the number of variables or the number of constraints that are needed to represent and formulate the complexities of the particular problem under consideration. For formulations that are out of reach of the capability of modern solvers, one strategy is to employ a *divide-and-conquer* strategy, whereby the formulation is reduced to smaller and easier-to-solve subproblems. The caveat with such a strategy is that one needs to be able to piece together the individual solutions of smaller subproblems in such a way so as to be able to obtain an optimal solution to the original model. This section will describe two algorithms that can be used for this purpose, namely Benders decomposition and Lagrangean relaxation. For ease of illustration, we will restrict the exposition to solving *linear* mixed-integer programming formulations. For extensions of these techniques to non-linear programming formulations, the reader may consult the references provided at the end of the chapter.

The formulation that will be used in the illustration of the techniques is assumed to take the following general form:

$$(\mathcal{M}) \quad \text{Minimise } cx + fy$$

subject to

$$Ax + By = d$$
$$x \in X$$
$$y \in Y,$$

where

    x and y are the column vectors of variables
    c and f are the row vectors of cost coefficients
    A and B are the constraint coefficient matrices
    d is the column vector of right-hand-side values, all with appropriate dimensions
    X and Y are non-empty sets in which variables x and y are defined, respectively

8.2.1 Benders decomposition

Benders decomposition (Benders, 1962) is a technique that partitions the model into two sets of variables, where one set induces smaller formulation called the *subproblem*, and the other set forms what is commonly named the *master* problem. There may be different ways of partitioning the variables of the original formulation, which in turn would give rise to different subproblems and master problems. Often, the formulation itself suggests a natural partitioning of the variables, such as x and y in model $\mathcal{M}$. In linear programming formulations, the partitioning of the variables can be made to correspond to two separate sets of decisions they represent. In mixed-integer linear programming formulations, it is customary to partition the variables such that the integer variables are separated from their continuous counterpart. However, these are not requirements but merely rules of thumb.

Benders decomposition seeks to identify and temporarily 'eliminate' a complicating set of variables from the formulation. The complicating set of variables can be regarded as those which, if removed from the formulation, would result in a subproblem that is much easier to solve in comparison to the original formulation. There may be several sets of such variables.

To illustrate the application of Benders' decomposition on formulation $\mathcal{M}$, we will first assume that variables x are continuous. We make no assumptions on the nature of the variables y, which can either be continuous or integer. It is possible to rewrite formulation $\mathcal{M}$ in the following form:

$$\text{Minimise } fy + \{\underset{x \in X}{\text{Minimise}}\{cx : Ax = d - By\}\}$$

subject to

$$y \in Y.$$

The inner minimisation problem is the primal subproblem referred here, which will be parametrised on variables y and shown as $S(y)$. As the subproblem $S(y)$ is linear and expressed in terms of the continuous variables x, it is possible to replace it with its dual $S'(y)$, as shown in the following:

$$(\mathcal{M}') \qquad \text{Minimise } fy + \{\underset{w}{\text{Maximise}}\{w(d - By) : wA \le c\}\}$$

subject to

$$y \in Y,$$

where w are the dual variables. Formulation $\mathcal{M}'$ is of a *min-max* type, which can be reformulated by using an auxiliary variable z corresponding to the cost of the subproblem. We now investigate what information can be obtained from the subproblem for a given fixing $y = \bar{y}$ of the variables.

There are two possible cases in relation to the solution of the subproblem:

1. $S(\bar{y})$ is infeasible. In other words, the fixing $\bar{y}$ does not yield a feasible solution for the original formulation $\mathcal{M}$. In this case, $S(\bar{y})$ will provide an extreme ray w'. It is then possible to use the following constraint to forbid the infeasibility,

$$w'(d - By) \leq 0, \tag{8.1}$$

referred to as an *infeasibility cut*.
2. $S(\bar{y})$ is feasible. In this case, it will yield an optimal dual solution w^* and an optimal value $w^*(d - By)$, which can be used to construct the following inequality:

$$z \geq w^*(d - By), \tag{8.2}$$

referred to as an *optimality cut*, and will provide a lower bound to the original problem using the optimal value of the subproblem.

Using the two sets of cuts, it is possible to reformulate $\mathcal{M}'$ further in the following form:

$(\mathcal{MP})$ Minimise $fy + z$

subject to

$$w(d - By) \leq 0 \qquad\qquad w \in W_D$$
$$z \geq w(d - By) \quad w \in W_P$$
$$y \in Y,$$

which is the master problem that was referred to earlier. The $\mathcal{MP}$ includes feasibility cuts written for each extreme ray in the set W_D of extreme rays of the dual of the subproblem, whenever it is infeasible. Similarly, it includes the optimality cuts for each solution w in the set W_P of extreme points of the feasible space of the dual subproblem, whenever it is feasible. These two sets of constraints are exponential in their number, for which reason it is impractical to solve the master problem in its initial form using off-the-shelf optimisers. Instead, it is more convenient to use a constraint generation algorithm, for which a pseudocode is given here.

Algorithm 8.1 Cutting plane algorithm

1: Set $t = 0$, $W_D^t = W_P^t = \emptyset$.
2: Set the upper bound $UB = \infty$ and the lower bound $LB = -\infty$.
3: **while** $UB - LB > \epsilon$ **do**
4:     Solve the relaxed master problem $\mathcal{MP}^t$ where $W_D = W_D^t$ and $W_P = W_P^t$.
5:     Set $LB = v(\mathcal{MP}^t)$.
6:     **if** the $\mathcal{MP}^t$ is infeasible **then**
7:         Stop.
8:     **else**
9:         Solve $S'(\bar{y})$ where $\bar{y}$ is an optimal solution of $\mathcal{MP}^t$ and $v(S'(\bar{y}))$ is the optimal value.
10:         **if** $S'(\bar{y})$ is unbounded **then**
11:             $W_D^{t+1} = W_D^{t+1} \cup \{w\}$, where w is an unbounded extreme ray.
12:         **else**
13:             $W_P^{t+1} = W_P^{t+1} \cup \{w\}$, where w is an optimal extreme point.
14:             If $f\bar{y} + v(S'(\bar{y})) < UB$ set $UB = f\bar{y} + v(S'(\bar{y}))$.
15:         **end if**
16:     **end if**
17:     Set $t = t + 1$.
18: **end while**

Several notes here are in order:

- The subproblem can sometimes be decomposed further into a collection of smaller subproblems. In this case, each smaller problem can be solved independently and will induce either a feasibility or an optimality cut.
- Benders decomposition can be applied to linear or integer linear programming formulations. It can also be extended to solve non-linear counterparts of such formulations which will not be covered here. The interested reader is referred to Geoffrion (1972) for more details on the latter.

8.2.2 Application to uncapacitated network design

We illustrate the application of Benders decomposition to the uncapacitated network design problem, using two different formulations of the problem.

8.2.2.1 Benders decomposition using an aggregated formulation

Using the parameters and variables defined in Chapter 3, we use the following formulation of the uncapacitated network design problem and refer to it as UND_a:

$$\text{Minimise} \sum_{(i,j)\in A} g_{ij}y_{ij} + \sum_{(i,j)\in A} \sum_{p\in P} c_{ij}^p x_{ij}^p$$

subject to

$$\sum_{j\in V_i^+} x_{ij}^p - \sum_{j\in V_i^-} x_{ji}^p = \begin{cases} 1, & \text{if } i = o(p) \\ -1, & \text{if } i = d(p) \\ 0, & \text{otherwise.} \end{cases} \quad \forall i \in V, p \in P$$

$$\sum_{p\in P} x_{ij}^p \leq |P|y_{ij} \quad \forall (i,j) \in A$$

$$x_{ij}^p \geq 0 \quad \forall (i,j) \in A, p \in P$$

$$y_{ij} \in \{0,1\} \quad \forall (i,j) \in A.$$

The formulation consists of two sets of constraints, where the first guarantee that each commodity is sent from its origin to destination, and the second set models the logical implication that for any amount of positive flow on a given arc $(i,j) \in P$, the binary variable takes the value 1 and activates the fixed cost g_{ij} in the objective function. The reason for referring to this model as an *aggregate* formulation stems from the fact that the latter set of constraints, which are also known as linking constraints, aggregate all flows on a given arc.

Formulation UND$_a$ suggests a natural partitioning of the variables into two sets, one relevant to the design (y variables) and the other pertaining to flow (x variables). Fixing the design variables as $y = \bar{y}$ induces the following subproblem, which we denote as SP$_a(\bar{y})$:

$$\text{Minimise} \sum_{(i,j)\in A} \sum_{p\in P} c_{ij}^p x_{ij}^p$$

subject to

$$\sum_{j\in V_i^+} x_{ij}^p - \sum_{j\in V_i^-} x_{ji}^p = \begin{cases} 1, & \text{if } i = o(p) \\ -1, & \text{if } i = d(p) \\ 0, & \text{otherwise.} \end{cases} \quad \forall i \in V, p \in P \tag{8.3}$$

$$-\sum_{p\in P} x_{ij}^p \geq -|P|\bar{y}_{ij} \quad \forall (i,j) \in A \tag{8.4}$$

$$x_{ij}^p \geq 0 \quad \forall (i,j) \in A, p \in P.$$

Defining dual variables α_i^p corresponding to constraints (8.3) and $\beta_{ij} \geq 0$ to constraints (8.4) results in the following dual of the subproblem named as $DSP_a(\bar{y})$:

$$\text{Maximise} \sum_{p \in P} \alpha_{o(p)}^p - \sum_{p \in P} \alpha_{d(p)}^p - \sum_{(i,j) \in A} |P| \bar{y}_{ij} \beta_{ij}$$

subject to

$$\alpha_i^p - \alpha_j^p - \beta_{ij} \leq c_{ij}^p \quad \forall (i,j) \in A, p \in P \tag{8.5}$$

$$\beta_{ij} \geq 0 \quad \forall (i,j) \in A.$$

The master problem as a reformulation of the uncapacitated network design problem can now be presented as follows:

$$\text{Minimise} \sum_{(i,j) \in A} g_{ij} y_{ij} + z$$

subject to

$$\sum_{p \in P} \bar{\alpha}_{o(p)}^p - \sum_{p \in P} \bar{\alpha}_{d(p)}^p - \sum_{(i,j) \in A} |P| \bar{\beta}_{ij} y_{ij} \leq 0 \quad (\bar{\alpha}, \bar{\beta}) \in W_D$$

$$\sum_{p \in P} \bar{\alpha}_{o(p)}^p - \sum_{p \in P} \bar{\alpha}_{d(p)}^p - \sum_{(i,j) \in A} |P| \bar{\beta}_{ij} y_{ij} \leq z \quad (\bar{\alpha}, \bar{\beta}) \in W_P$$

$$y_{ij} \in \{0, 1\} \quad \forall (i,j) \in A.$$

where the first set of constraints are the feasibility cuts written for set W_D of all extreme rays $(\bar{\alpha}, \bar{\beta})$ of the dual subproblem when it is unbounded, indicating that the primal subproblem is infeasible. In contrast, if the primal subproblem is feasible, then there exists an extreme point $(\bar{\alpha}, \bar{\beta})$ in the set W_P of feasible solutions to the dual subproblem, inducing an optimality cut as per the second set of constraints in the master problem.

We illustrate the application of the algorithm on the following numerical example.

Example 8.1

Consider a four-node network as shown in Figure 8.1 that will be used to ship two commodities. A single unit of commodity p_1 originates from node 1 and is destined to node 3. A single unit of commodity p_2 originates from node 1 and is destined to node 4. We assume a unit variable cost of shipment does not vary from one commodity to the other

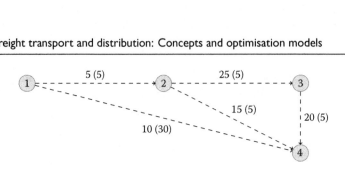

Figure 8.1 The four-node network used for the uncapacitated network design instance.

(i.e. $c_{ij}^{p_1} = c_{ij}^{p_2}$). The figure also shows, next to each arc, the fixed cost and the variable cost (the latter in parentheses) of that arc.

The following is formulation UND$_a$ as applied to the numerical example.

$$\text{Minimise } 5y_{12} + 10y_{14} + 25y_{23} + 15y_{24} + 20y_{34}$$

$$+5x_{12}^1 + 30x_{14}^1 + 5x_{23}^1 + 5x_{24}^1 + 5x_{34}^1$$

$$+5x_{12}^2 + 30x_{14}^2 + 5x_{23}^2 + 5x_{24}^2 + 5x_{34}^2$$

subject to

$$x_{12}^1 = 1$$

$$x_{23}^1 + x_{24}^1 - x_{12}^1 = 0$$

$$x_{34}^1 - x_{23}^1 = -1$$

$$-x_{14}^1 - x_{24}^1 - x_{34}^1 = 0$$

$$x_{12}^2 = 1$$

$$x_{23}^2 + x_{24}^2 - x_{12}^2 = 0$$

$$x_{34}^2 - x_{23}^2 = 0$$

$$-x_{14}^2 - x_{24}^2 - x_{34}^2 = -1$$

$$-x_{12}^1 - x_{12}^2 \geq -2y_{12}$$

$$-x_{14}^1 - x_{14}^2 \geq -2y_{14}$$

$$-x_{23}^1 - x_{23}^2 \geq -2y_{23}$$

$$-x_{24}^1 - x_{24}^2 \geq -2y_{24}$$

$$-x_{34}^1 - x_{34}^2 \geq -2y_{34}$$

$$\text{All } x_{ij}^p \geq 0$$

$$\text{All } y_{ij} \in \{0, 1\}.$$

For a given fixing $\bar{y}$ of the design variables, the dual subproblem DSP_a for this instance takes the following form:

Maximise $\alpha_1^1 - \alpha_3^1 + \alpha_4^2 - \alpha_4^2$

$$-2\bar{y}_{12}\beta_{12} - 2\bar{y}_{14}\beta_{14} - 2\bar{y}_{23}\beta_{23} - 2\bar{y}_{24}\beta_{24} - 2\bar{y}_{34}\beta_{34}$$

subject to

$$\alpha_1^1 - \alpha_2^1 - \beta_{12} \leq 5$$

$$\alpha_1^1 - \alpha_4^1 - \beta_{14} \leq 30$$

$$\alpha_2^1 - \alpha_3^1 - \beta_{23} \leq 5$$

$$\alpha_2^1 - \alpha_4^1 - \beta_{24} \leq 5$$

$$\alpha_3^1 - \alpha_4^1 - \beta_{34} \leq 5$$

$$\alpha_1^2 - \alpha_2^2 - \beta_{12} \leq 5$$

$$\alpha_1^2 - \alpha_4^2 - \beta_{14} \leq 30$$

$$\alpha_2^2 - \alpha_3^2 - \beta_{23} \leq 5$$

$$\alpha_2^2 - \alpha_4^2 - \beta_{24} \leq 5$$

$$\alpha_3^2 - \alpha_4^2 - \beta_{34} \leq 5$$

$$\text{All } \beta_{ij} \geq 0.$$

We now present a step-by-step application of the constraint generation method as shown in Algorithm 8.1.

0. Initialise $UB = \infty$, $LB = -\infty$. In representing the solutions, we will use the following vectors:

- $\bar{y} = (\bar{y}_{12}, \bar{y}_{14}, \bar{y}_{23}, \bar{y}_{24}, \bar{y}_{34})$,
- $\bar{\alpha} = (\bar{\alpha}_1^1, \bar{\alpha}_2^1, \bar{\alpha}_3^1, \bar{\alpha}_4^1, \bar{\alpha}_1^2, \bar{\alpha}_2^2, \bar{\alpha}_3^2, \bar{\alpha}_4^2)$,
- $\bar{\beta} = (\bar{\beta}_{12}, \bar{\beta}_{14}, \bar{\beta}_{23}, \bar{\beta}_{24}, \bar{\beta}_{34})$.

1. Solve the following master problem:

Minimise $5y_{12} + 10y_{14} + 25y_{23} + 15y_{24} + 20y_{34} + z$

subject to

$$\text{All } y_{ij} \in \{0, 1\}.$$

which yields the optimal value 0 corresponding to the solution vector $\bar{y} = (0, 0, 0, 0, 0)$ and $\bar{z} = 0$. In this case, $SP_a(\bar{y})$ is infeasible as there are no paths that would allow either commodity to reach their destination, for which the solution of the $DSP_a(\bar{y})$ yields the following extreme ray $(\bar{\alpha}, \bar{\beta}) = (3, 2, 1, 2, 2, 1, 1, 1, 1, 1, 1, 0, 0)$, implying the following feasibility cut:

$$y_{12} + y_{14} + y_{23} \geq 3/2,$$

which has the intuitive interpretation that at least two of the arcs in the set $\{(1, 2), (1, 4), (2, 3)\}$ should be open. The cut is added to the master problem.

2. $UB = \infty$, $LB = 0$. The augmented master problem, namely

$$\text{Minimise } 5y_{12} + 10y_{14} + 25y_{23} + 15y_{24} + 20y_{34} + z$$

subject to

$$y_{12} + y_{14} + y_{23} \geq 3/2$$
$$\text{All } y_{ij} \in \{0, 1\},$$

has the optimal solution $\bar{y} = (1, 1, 0, 0, 0)$ that satisfies the new constraint, yielding $\bar{z} = 15$. $SP_a(\bar{y})$ in this case is still infeasible, as the active arcs in the network do not provide the connectivity needed to send commodity p_1 from its origin node 1 to its destination node 3. An extreme ray $(\bar{\alpha}, \bar{\beta}) = (2, 2, 1, 2, 1, 1, 1, 1, 0, 0, 1, 0, 0)$ obtained from the $DSP_a(\bar{y})$ corresponds to the following feasibility cut:

$$y_{23} \geq 1/2.$$

The natural interpretation of this cut is that arc $(2, 3)$ must be in the set of cuts selected so as to be able to send commodity p_1 from its origin to destination. The cut is appended to the master problem for the next iteration.

3. $UB = \infty$, $LB = 15$. The augmented master problem, namely

$$\text{Minimise } 5y_{12} + 10y_{14} + 25y_{23} + 15y_{24} + 20y_{34} + z$$

subject to

$$y_{12} + y_{14} + y_{23} \geq 3/2$$
$$y_{23} \geq 1/2$$
$$\text{All } y_{ij} \in \{0, 1\},$$

has the optimal solution $\bar{y} = (1, 0, 1, 0, 0)$ with value 30, which again does not include a sufficient number of arcs to be able to ship commodity p_2 to its destination. For this reason, $DSP_a(\bar{y})$ is unbounded and returns an extreme ray $(\bar{\alpha}, \bar{\beta}) = (1, 1, 1, 0, 2, 2, 2, 1, 0, 1, 0, 1, 1)$, returning the following feasibility cut:

$$y_{14} + y_{24} + y_{34} \geq 1/2,$$

which ensures that there is at least one arc leading into node 4 as the destination of commodity p_2.

4. $UB = \infty$, $LB = 30$. The augmented master problem is as follows:

$$\text{Minimise } 5y_{12} + 10y_{14} + 25y_{23} + 15y_{24} + 20y_{34} + z$$

subject to

$$y_{12} + y_{14} + y_{23} \geq 3/2$$
$$y_{23} \geq 1/2$$
$$y_{14} + y_{24} + y_{34} \geq 1/2$$
$$\text{All } y_{ij} \in \{0, 1\},$$

which yields the optimal solution $\bar{y} = (0, 1, 1, 0, 0)$ with value 35, which still is infeasible in terms of the routing subproblem, but this time in terms of ensuring connectivity for p_1. In this case, an extreme ray $(\bar{\alpha}, \bar{\beta}) = (2, 1, 1, 2, 1, 1, 1, 1, 1, 0, 0, 0, 0)$ is obtained from the $DSP_a(\bar{y})$, which in turn is to construct the following feasibility cut:

$$y_{12} \geq 1/2.$$

5. $UB = \infty$, $LB = 35$. The master problem at this iteration includes four feasibility cuts, as shown here:

$$\text{Minimise } 5y_{12} + 10y_{14} + 25y_{23} + 15y_{24} + 20y_{34} + z$$

subject to

$$y_{12} + y_{14} + y_{23} \geq 3/2$$
$$y_{23} \geq 1/2$$
$$y_{14} + y_{24} + y_{34} \geq 1/2$$
$$y_{12} \geq 1/2$$
$$\text{All } y_{ij} \in \{0, 1\},$$

which has the optimal solution $\bar{y} = (1, 1, 1, 0, 0)$. It is interesting to see that the feasibility cuts generated so far correspond to connectivity constraints to ultimately ensure that, for each commodity, there exists a path between the origin and the destination.

The solution of the master problem in this iteration now implies a feasible subproblem $SP_a(\bar{y})$, with a value equal to 40. Together with the fixed cost of the arcs $(1, 2)$, $(1, 4)$ and $(2, 3)$, this yields an upper bound $UB = 80$. By solving the $DSP_a(\bar{y})$, one obtains the extreme point $(\bar{\alpha}, \bar{\beta}) = (10, 5, 0, 0, 0, 0, 0, -30, 0, 0, 0, 25, 25)$. If we take a closer look at this solution, it can be seen that the values are such that $\bar{\alpha}_1^1 - \bar{\alpha}_3^1 = 10$ is the routing cost for commodity p_1 with respect to the active arcs $(1, 2)$ and $(1, 4)$. Similarly, $\bar{\alpha}_1^2 - \bar{\alpha}_4^1 = 30$ is the routing cost for commodity p_2, flowing on arc $(1, 4)$. More interestingly, the dual variables $\bar{\beta}_{24} = \bar{\beta}_{34} = 25$ show the *potential* decrease in the routing cost if we were to use either arc corresponding to these two variables. Indeed, moving the flow of commodity p_2 from arc $(1, 4)$ to arc $(2, 4)$ results in a cost difference of $30 - 5 = 25$ units. Using the dual optimal solution, an optimality cut can be constructed as

$$40 - 50y_{24} - 50y_{34} \geq z,$$

which is added to the master problem. It is noteworthy that the optimality cut applies the potential reduction in cost to both commodities as 2×25, for each arc. This is an overestimation which stems from using an aggregated formulation.

6. $UB = 80$, $LB = 40$. The master problem with the new optimality cut reads as follows:

$$\text{Minimise } 5y_{12} + 10y_{14} + 25y_{23} + 15y_{24} + 20y_{34} + z$$

subject to

$$y_{12} + y_{14} + y_{23} \geq 3/2$$
$$y_{23} \geq 1/2$$
$$y_{14} + y_{24} + y_{34} \geq 1/2$$
$$y_{12} \geq 1/2$$
$$40 - 50y_{24} - 50y_{34} \geq z$$
$$\text{All } y_{ij} \in \{0, 1\},$$

which has the optimal solution $\bar{y} = (1, 0, 1, 1, 0)$, $\bar{z} = 0$ and an optimal value 45. The associated routing problem is the solution of $SP_a(\bar{y})$, with value 20, implies a new $UB = 65$ and returns the extreme point $(\bar{\alpha}, \bar{\beta}) = (10, 5, 0, 0, 5, 0, 0, -5, 0, 0, 0, 0, 0)$. All dual variables $\bar{\beta}$ are zero, yielding the following optimality cut:

$$20 \geq z.$$

It is easy to show by enumeration that the minimum possible routing cost that can be achieved for this instance is 20, and this optimality cut implies exactly this.

7. $UB = 65$, $LB = 45$. Adding the latest optimality cut to the master problem results in the formulation here:

$$\text{Minimise } 5y_{12} + 10y_{14} + 25y_{23} + 15y_{24} + 20y_{34} + z$$

subject to

$$y_{12} + y_{14} + y_{23} \geq 3/2$$
$$y_{23} \geq 1/2$$
$$y_{14} + y_{24} + y_{34} \geq 1/2$$
$$y_{12} \geq 1/2$$
$$40 - 50y_{24} - 50y_{34} \geq z$$
$$20 \geq z$$
$$\text{All } y_{ij} \in \{0, 1\},$$

which has the optimal solution $\bar{y} = (1, 0, 1, 1, 0)$, $\bar{z} = 20$ and value 65. Since $UB - LB = 0$, the algorithm stops, returning $\bar{y} = (1, 0, 1, 1, 0)$ as the optimal solution to the overall problem.

The application of Benders decomposition to the instance in Figure 8.1 results in a master problem that has a total of six constraints, as opposed to the original formulation of the problem with 13 constraints. If the subproblems can be solved efficiently, the decomposition algorithm in most instances will prove to be faster than solving the original formulation, particularly for larger-size networks with an increased number of commodities.

We now look at using a slightly different formulation of the uncapacitated network design problem, and the effect of this formulation on the application of Benders decomposition.

8.2.2.2 Benders decomposition using a stronger formulation

An alternative formulation of the uncapacitated network design problem, which we refer to as UND$_s$, is presented as follows:

$$\text{Minimise} \sum_{(i,j)\in A} g_{ij}y_{ij} + \sum_{(i,j)\in A}\sum_{p\in P} c_{ij}^p x_{ij}^p$$

subject to

$$\sum_{j\in V_i^+} x_{ij}^p - \sum_{j\in V_i^-} x_{ji}^p = \begin{cases} 1, & \text{if } i = o(p) \\ -1, & \text{if } i = d(p) \\ 0, & \text{otherwise.} \end{cases} \quad \forall i \in V, p \in P$$

$$x_{ij}^p \leq y_{ij} \qquad\qquad \forall (i,j) \in A, p \in P$$

$$x_{ij}^p \geq 0 \qquad\qquad \forall (i,j) \in A, p \in P$$

$$y_{ij} \subset \{0,1\} \qquad\qquad \forall (i,j) \in A.$$

Formulation UND_s uses a slightly different set of linking constraints between defined for each arc (i,j) and commodity p separately. It is easy to see that the linking constraints in UND_s are disaggregations of those in UND_a and imply the latter when aggregated over all commodities $p \in P$. This observation can be used to show that the LP relaxation of UND_s is at least as good as that of UND_a and is therefore said to be a *stronger* formulation.

To see the effect of using such disaggregated linking constraints and a stronger formulation on the application of Benders decomposition, we fix the design variables in UND_s in the same way as for UND_s. In particular, when $y = \bar{y}$, the formulation yields a subproblem. However, this time the subproblem decomposes into smaller subproblems, one for each commodity $p \in P$, denoted $SP_s^p(\bar{y})$, and takes the following form:

$$\text{Minimise} \sum_{(i,j)\in A} c_{ij}^p x_{ij}^p$$

subject to

$$\sum_{j\in V_i^+} x_{ij}^p - \sum_{j\in V_i^-} x_{ji}^p = \begin{cases} 1, & \text{if } i = o(p) \\ -1, & \text{if } i = d(p) \\ 0, & \text{otherwise.} \end{cases} \quad \forall i \in V \qquad (8.6)$$

$$x_{ij}^p \leq \bar{y}_{ij} \qquad\qquad \forall (i,j) \in A \qquad (8.7)$$

$$x_{ij}^p \geq 0 \qquad\qquad \forall (i,j) \in A.$$

It is easy to see that $SP_s^p(\bar{y})$ is a shortest-path problem for commodity $p \in P$ on a restricted network, the solution of which will satisfy one of the two cases here:

1. $SP_s^p(\bar{y})$ is feasible, in which case it will yield the shortest path for the commodity from $o(p)$ to $o(d)$. In this case, if α and β are the dual variables corresponding to constraints (8.6) and (8.7), an extreme point $(\bar{\alpha}^p, \bar{\beta}^p)$ obtained by solving SP_s^p will yield the following optimality cut:

$$\bar{\alpha}_{o(p)}^p - \bar{\alpha}_{d(p)}^p - \sum_{(i,j)\in A} \bar{\beta}_{ij}^p y_{ij} \leq z^p,$$

where z^p is a lower bound on the value of the subproblem $p \in P$.

2. $SP_s^p(\bar{y})$ is infeasible, which indicates that there does not exist a path from $o(p)$ to $d(p)$ on which commodity p can be sent, and will in turn yield the following feasibility cut:

$$\bar{\alpha}_{o(p)}^p - \bar{\alpha}_{d(p)}^p - \sum_{(i,j)\in A} \bar{\beta}_{ij}^p y_{ij} \leq 0,$$

where $(\bar{\alpha}, \bar{\beta})$ is an extreme ray obtained by solving the dual of $SP_s^p(\bar{y})$.

The reformulation of the stronger formulation as the Benders master problem can now be written as follows:

$$\text{Minimise } \sum_{(i,j)\in A} g_{ij} y_{ij} + \sum_{p\in P} z^p$$

subject to

$$\bar{\alpha}_{o(p)}^p - \bar{\alpha}_{d(p)}^p - \sum_{(i,j)\in A} \bar{\beta}_{ij}^p y_{ij} \leq 0 \qquad (\bar{\alpha}, \bar{\beta}) \in W_D$$

$$\bar{\alpha}_{o(p)}^p - \bar{\alpha}_{d(p)}^p - \sum_{(i,j)\in A} \bar{\beta}_{ij}^p y_{ij} \leq z^p \qquad (\bar{\alpha}, \bar{\beta}) \in W_P$$

$$y_{ij} \in \{0, 1\} \quad \forall (i,j) \in A.$$

Example 8.2

This example presents an application of Benders decomposition using the strong formulation for the numerical instance shown in Figure 8.1. The formulation itself is shown in the following:

$$\text{Minimise } 5y_{12} + 10y_{14} + 25y_{23} + 15y_{24} + 20y_{34}$$
$$+5x_{12}^1 + 30x_{14}^1 + 5x_{23}^1 + 5x_{24}^1 + 5x_{34}^1$$
$$+5x_{12}^2 + 30x_{14}^2 + 5x_{23}^2 + 5x_{24}^2 + 5x_{34}^2$$

subject to

$$x_{12}^1 = 1$$

$$x_{23}^1 + x_{24}^1 - x_{12}^1 = 0$$

$$x_{34}^1 - x_{23}^1 = -1$$

$$-x_{14}^1 - x_{24}^1 - x_{34}^1 = 0$$

$$x_{12}^2 = 1$$

$$x_{23}^2 + x_{24}^2 - x_{12}^2 = 0$$

$$x_{34}^2 - x_{23}^2 = 0$$

$$-x_{14}^2 - x_{24}^2 - x_{34}^2 = -1$$

$$-x_{12}^1 \geq -y_{12}$$

$$-x_{14}^1 \geq -y_{14}$$

$$-x_{23}^1 \geq -y_{23}$$

$$-x_{24}^1 \geq -y_{24}$$

$$-x_{34}^1 \geq -y_{34}$$

$$-x_{12}^2 \geq -y_{12}$$

$$-x_{14}^2 \geq -y_{14}$$

$$-x_{23}^2 \geq -y_{23}$$

$$-x_{24}^2 \geq -y_{24}$$

$$-x_{34}^2 \geq -y_{34}$$

$$\text{All } x_{ij}^p \geq 0$$

$$\text{All } y_{ij} \in \{0, 1\}.$$

For a given fixing $\bar{y}$ of the design variables, the dual subproblem decomposes into two, one for p_1, denoted $\mathrm{DSP}_a^{p_1}$, as shown here:

$$(\mathrm{DSP}_a^{p_1}) \text{ Maximise } \alpha_1^1 - \alpha_3^1$$

$$-\bar{y}_{12}\beta_{12}^1 - \bar{y}_{14}\beta_{14}^1 - \bar{y}_{23}\beta_{23}^1 - \bar{y}_{24}\beta_{24}^1 - \bar{y}_{34}\beta_{34}^1$$

subject to

$$\alpha_1^1 - \alpha_2^1 - \beta_{12}^1 \leq 5$$

$$\alpha_1^1 - \alpha_4^1 - \beta_{14}^1 \leq 30$$

$$\alpha_2^1 - \alpha_3^1 - \beta_{23}^1 \leq 5$$

$$\alpha_2^1 - \alpha_4^1 - \beta_{24}^1 \leq 5$$

$$\alpha_3^1 - \alpha_4^1 - \beta_{34}^1 \leq 5$$

$$\text{All } \beta_{ij} \geq 0,$$

and the other for p_2, denoted $\text{DSP}_a^{p_2}$, presented here:

$$(\text{DSP}_a^{p_2}) \text{ Maximise } \alpha_4^2 - \alpha_4^2$$

$$-\bar{y}_{12}\beta_{12}^2 - \bar{y}_{14}\beta_{14}^2 - \bar{y}_{23}\beta_{23}^2 - \bar{y}_{24}\beta_{24}^2 - \bar{y}_{34}\beta_{34}^2$$

subject to

$$\alpha_1^2 - \alpha_2^2 - \beta_{12}^2 \leq 5$$

$$\alpha_1^2 - \alpha_4^2 - \beta_{14}^2 \leq 30$$

$$\alpha_2^2 - \alpha_3^2 - \beta_{23}^2 \leq 5$$

$$\alpha_2^2 - \alpha_4^2 - \beta_{24}^2 \leq 5$$

$$\alpha_3^2 - \alpha_4^2 - \beta_{34}^2 \leq 5$$

$$\text{All } \beta_{ij} \geq 0.$$

0. Initialise $UB = \infty$, $LB = -\infty$. In this example, we will use the following vectors to represent the solutions.

- $\bar{y} = (\bar{y}_{12}, \bar{y}_{14}, \bar{y}_{23}, \bar{y}_{24}, \bar{y}_{34})$,
- $\bar{\alpha}^1 = (\bar{\alpha}_1^1, \bar{\alpha}_2^1, \bar{\alpha}_3^1, \bar{\alpha}_4^1)$,
- $\bar{\alpha}^2 = (\bar{\alpha}_1^2, \bar{\alpha}_2^2, \bar{\alpha}_3^2, \bar{\alpha}_4^2)$,
- $\bar{\beta}^1 = (\bar{\beta}_{12}^1, \bar{\beta}_{14}^1, \bar{\beta}_{23}^1, \bar{\beta}_{24}^1, \bar{\beta}_{34}^1)$.
- $\bar{\beta}^2 = (\bar{\beta}_{12}^2, \bar{\beta}_{14}^2, \bar{\beta}_{23}^2, \bar{\beta}_{24}^2, \bar{\beta}_{34}^2)$.

1. $UB = \infty$, $LB = -\infty$. The initial master problem, namely

$$\text{Minimise } 5y_{12} + 10y_{14} + 25y_{23} + 15y_{24} + 20y_{34} + z^1 + z^2$$

subject to

$$\text{All } y_{ij} \in \{0, 1\}.$$

has the optimal value 0 corresponding to the solution vector $\bar{y} = (0, 0, 0, 0, 0)$ and $\bar{z} = 0$. $\text{SP}_a(\bar{y})$ is clearly infeasible in this case. The extreme ray obtained from DSP_a^{p1} is $(\bar{\alpha}^1, \bar{\beta}^1) = (2, 1, 0, 2, 1, 0, 1, 0, 0)$, which induces the following feasibility cut:

$$y_{12} + y_{23} \geq 2,$$

which requires that *both* arcs $(1, 2)$ and $(2, 3)$ are to be active, thereby ensuring connectivity in the network for sending commodity p_1 from node 1 to node 3. On the other hand, the extreme ray extracted from DSP_a^{p2} is $(\bar{\alpha}^2, \bar{\beta}^2) = (2, 1, 1, 1, 1, 1, 0, 0, 0)$, which generates the following feasibility cut:

$$y_{12} + y_{14} \geq 1,$$

which corresponds to the two arcs separating the origin node 1 of commodity p_2 from the destination node 4. In other words, the feasibility cut implies that at least one of the two arcs $(1, 2)$ and $(1, 4)$ should be used in sending commodity p_2 from node 1.

2. $UB = \infty$, $LB = -\infty$. The master problem with the two additional feasibility cuts is presented here:

$$\text{Minimise } 5y_{12} + 10y_{14} + 25y_{23} + 15y_{24} + 20y_{34} + z^1 + z^2$$

subject to

$$y_{12} + y_{23} \geq 2$$
$$y_{12} + y_{14} \geq 1$$
$$\text{All } y_{ij} \in \{0, 1\},$$

has the optimal value 0 corresponding to the solution vector $\bar{y} = (1, 0, 1, 0, 0)$ and $\bar{z} = 30$. The routing subproblem for p_1 is now feasible, with an optimal value equal to 10, and in this case DSP_a^{p1} returns an extreme point $(\bar{\alpha}^1, \bar{\beta}^1) = (10, 5, 0, 0, 0, 0, 0, 0, 0)$, implying the following optimality cut:

$$10 \geq z^1,$$

which states that the minimum achievable cost of sending p_1 from its origin to destination is 10. The routing subproblem DSP_a^{p2} for p_2 returns an extreme ray $(\bar{\alpha}^2, \bar{\beta}^2) = (1, 1, 0, 0, 0, 1, 1, 1, 0)$, yielding the following feasibility cut:

$$y_{14} + y_{24} + y_{34} \geq 1,$$

which is required to ensure the connectivity in the network assuming that arc $(1, 2)$ is active. Both these cuts are added to the master problem for the following iteration.

3. $UB = \infty$, $LB = 30$. The master problem with the two additional feasibility cuts is presented here:

$$\text{Minimise } 5y_{12} + 10y_{14} + 25y_{23} + 15y_{24} + 20y_{34} + z^1 + z^2$$

subject to

$$y_{12} + y_{23} \geq 2$$
$$y_{12} + y_{14} \geq 1$$
$$10 \geq z^1$$
$$y_{14} + y_{24} + y_{34} \geq 1$$
$$\text{All } y_{ij} \in \{0, 1\}.$$

The new master problem returns the optimal value 50 and optimal solution $\bar{y} = (1, 1, 1, 0, 0)$. In this case, DSP_a^{p1} provides an extreme point that produces the same cut as mentioned earlier, namely $z^1 \geq 10$. On the other hand, DSP_a^{p2} is now bounded, has an optimal value equal to 30 and returns an extreme point $(\bar{\alpha}^2, \bar{\beta}^2) = (0, 0, 0, -30, 0, 0, 0, 25, 25)$. This extreme ray is similar to that obtained in the previous example, where the $\beta_{24}^2 = \beta_{34}^2 = 25$ indicate the potential saving in the routing cost that can be obtained if arcs $(2, 4)$ and $(3, 4)$ are used to ship the flow of commodity p_2. The extreme point gives rise to the following optimality cut:

$$30 - 25y_{24} - 25y_{34} \geq z^2.$$

The new cut is added to the master problem and the upper bound is updated as $UB = 80$.

4. $UB = 80$, $LB = 50$. The updated master problem is as follows:

$$\text{Minimise } 5y_{12} + 10y_{14} + 25y_{23} + 15y_{24} + 20y_{34} + z^1 + z^2$$

subject to

$$y_{12} + y_{23} \geq 2$$
$$y_{12} + y_{14} \geq 1$$
$$10 \geq z^1$$
$$y_{14} + y_{24} + y_{34} \geq 1$$
$$30 - 25y_{24} - 25y_{34} \geq z^2$$
$$\text{All } y_{ij} \in \{0, 1\},$$

which has the optimal value 60 corresponding to the solution $\bar{y} = (1, 0, 1, 1, 0)$. As before, the solution $\text{DSP}_a^{p_1}$ does not provide a new cut. The optimal value of $\text{DSP}_a^{p_2}$ is 10, which provides the cut $z^2 \geq 10$, indicating the minimum cost at which commodity p_2 can be shipped. The new cut is added to the master problem, and the upper bound is updated as $UB = 65$.

5. $UB = 65$, $LB = 60$. The updated master problem now reads as follows:

$$\text{Minimise } 5y_{12} + 10y_{14} + 25y_{23} + 15y_{24} + 20y_{34} + z^1 + z^2$$

subject to

$$y_{12} + y_{23} \geq 2$$
$$y_{12} + y_{14} \geq 1$$
$$10 \geq z^1$$
$$y_{14} + y_{24} + y_{34} \geq 1$$
$$30 - 25y_{24} - 25y_{34} \geq z$$
$$10 \geq z^2$$
$$\text{All } y_{ij} \in \{0, 1\}.$$

the optimal value of which is 65, corresponding to the optimal solution $\bar{y} = (1, 0, 1, 1, 0)$. The algorithm now stops since $UB = LB$, returning $\bar{y} = (1, 0, 1, 1, 0)$ as the optimal solution to the overall problem.

It is worth noting that the optimality cuts resulting from the application of Benders decomposition on UND_a are precisely those that can be obtained by aggregating those from UND_s. It should also be noted that in the application of Algorithm 8.1 to the former case, each iteration resulted in one feasibility or optimality cut, whereas in the latter case, each iteration produced $|P|$ cuts, each being either a feasibility or an optimality cut per commodity. However, the algorithm using the stronger formulation converged in five iterations, where feasibility was restored earlier on in the process, whereas the former algorithm required seven iterations and required more iterations to identify a feasible solution.

8.2.3 Lagrangean relaxation

In its basic form, Lagrangean relaxation is a technique used to calculate lower bounds for formulation $\mathcal{M}$, and an ideal application and implementation should achieve this in a relatively quick fashion. If $\mathcal{M}$ is a linear programming formulation, Lagrangean relaxation, in theory, would be able to identify the optimal value. On the other hand, if $\mathcal{M}$ is an integer programming formulation, either linear or non-linear, then there is no guarantee that the technique would be able to yield the optimal value. In this case, one would resort to Lagrangean heuristics to calculate upper bounds.

In contrast to Benders decomposition, Lagrangean relaxation seeks to identify and temporarily eliminate 'complicating' sets of constraints, which are those that, when removed from formulation $\mathcal{M}$, would result in a subproblem that is much easier to solve. The technique works by relaxing (or dualising) the complicating constraints. In particular, if we assume that $Ax + By = d$ in $\mathcal{M}$ are the constraints that complicate the formulation, they can be relaxed by using a vector μ, unrestricted in sign, of Lagrangean multipliers as follows:

$$(\mathcal{M}_\mu) \quad \text{Minimise} \quad cx + fy + \mu(Ax + By - d)$$

subject to

$$x \in X, y \in Y.$$

Sometimes, and as is the case for formulation $\mathcal{M}$ mentioned earlier, the relaxed problem $\mathcal{M}_\mu$ suggests an immediate decomposition into two subproblems, one being

$$\underset{x \in X}{\text{Minimise}} \ (c + \mu A)x,$$

defined only in x variables, and the other being

$$\underset{y \in Y}{\text{Minimise}} \ (f + \mu B)y,$$

that is defined only in y variables.

The optimal value $v(\mathcal{M}_\mu)$ of $\mathcal{M}_\mu$, for any given μ, is a lower bound on the optimal value of $\mathcal{M}$. To find the best possible lower bound, one has to solve the following piecewise linear concave optimisation problem:

$$\underset{\mu}{\text{Maximise}} \ \mathcal{M}_\mu,$$

usually referred to as the Lagrangean dual problem and can be solved by means of non-differentiable optimisation techniques. One of the more popular techniques used for this purpose is subgradient optimisation, which in essence is an iterative search algorithm, where, at each iteration, a step is taken in the direction of the subgradient $Ax + By - d$. The overall scheme of subgradient optimisation, in its most general form, is shown in Algorithm 8.2.

Algorithm 8.2 Subgradient optimisation

1: Initialise multipliers as μ^1 and the maximum number t_{max} of iterations.
2: Set the best lower bound as $LB = -\infty$ and the upper bound as $UB = \infty$.
3: Set $t = 1$.
4: **while** $(UB - LB)/UB \geq \epsilon$ or $t \leq t_{max}$ **do**
5:     Solve $\mathcal{M}_{\mu^t}$
6:     **if** $v(\mathcal{M}_{\mu^t}) > LB$ **then**
7:        $LB = v(\mathcal{M}_{\mu^t})$
8:     **end if**
9:     Update UB, if needed.
10:    Update multipliers as $\mu^{t+1} = \mu^t + s^t \sigma^t$, where μ^t is the vector of multipliers, σ^t is the subgradient vector and s^t is the step-length at iteration t. The step-length s^t is calculated as follows:

$$s^t = \lambda \frac{UB - v(\mathcal{M}_{\mu^t})}{\|\sigma^t\|^2}.$$

11:    Set $t = t + 1$.
12: **end while**

The subgradient optimisation algorithm requires an upper bound to be able to calculate the step-length, as indicated in Steps 9 and 10. There are two ways of achieving this. The first is to use a relatively simple method, for example a heuristic (see the following section), to produce an upper bound and use this value as a constant throughout the algorithm. A second and a preferred way is to dynamically calculate an upper value based on the solution of the relaxed Lagrangean problem. The solution to $\mathcal{M}_\mu$ will generally be infeasible for $\mathcal{M}$, but it could be *repaired* using simple and problem-specific methods. Such a way of generating upper bounds for the problem is known as a *Lagrangean heuristic*. The upper bounds found in this way, however, are not guaranteed to be monotonically decreasing as one iterates within the algorithm.

Some further remarks on Lagrangean relaxation are as follows:

- Subgradient optimisation will typically require a large number of iterations to converge to satisfactory lower bounds. In this case, the efficiency of Lagrangean relaxation will heavily depend on the speed with which the subproblems can be solved to optimality. For this reason, when choosing the constraints to relax, one must ensure that the resulting subproblems will be much easier to solve than the original problem. If the subproblems have the same level of difficulty or complexity as the original problem, then the relaxation cannot be considered as efficient.

- Solving the Lagrangean dual involves optimising the values of the dual variables to achieve the tightest possible lower bound. If too many constraints are relaxed, this increases the difficulty of solving the Lagrangean dual problem as there will be as many dual variables as the number of constraints relaxed.
- If $\mathcal{M}$ is a (mixed-)integer programming formulation, then whether or not $\mathcal{M}$ has the integrality property will affect the strength of the Lagrangean dual. If $\mathcal{M}$ has the integrality property, then the strongest lower bound provided by the Lagrangean dual cannot be better than the optimal value of the linear programming relaxation of $\mathcal{M}$. Otherwise, the optimal value of the Lagrangean dual will be at least as good as the optimal value of the linear programming relaxation of $\mathcal{M}$.

8.2.4 Application to uncapacitated network design

This section illustrates the application of Lagrangean relaxation on the capacitated network design problem, introduced in Chapter 3. Using the same notation defined therein, the formulation of the problem we will work with is as follows:

$$\text{Minimise} \sum_{(i,j)\in A} g_{ij}y_{ij} + \sum_{(i,j)\in A}\sum_{p\in P} c_{ij}^p x_{ij}^p \tag{8.8}$$

subject to

$$\sum_{j\in V_i^+} x_{ij}^p - \sum_{j\in V_i^-} x_{ji}^p = d_i^p \qquad \forall i \in V, p \in P \tag{8.9}$$

$$\sum_{p\in P} x_{ij}^p \leq u_{ij}y_{ij} \quad \forall (i,j) \in A \tag{8.10}$$

$$x_{ij}^p \geq 0 \qquad \forall (i,j) \in A, p \in P \tag{8.11}$$

$$y_{ij} \in \{0,1\} \quad \forall (i,j) \in A. \tag{8.12}$$

In this formulation, constraints (8.9) ensure balance of flow in the network and that there is flow from the origin to destination of each commodity $p \in P$ in the amount d_i^p as defined in Chapter 3. Constraints (8.10) ensure that the total flow on each arc $(i,j) \in A$ in the network does not exceed the arc capacity u_{ij}.

This formulation suggests two different sets of constraints, one pertaining to flow and the other to capacity, suggesting that they could individually be relaxed in a Lagrangean fashion as detailed in the following:

8.2.4.1 Relaxing the arc constraints

We relax constraints (8.10) in a Lagrangean fashion by associating variables $\beta_{ij} \geq 0$ to these constraints, respectively. The resulting formulation of the relaxed problem can be stated as follows:

$$(LR_1(\beta)) \quad \text{Minimise} \sum_{(i,j) \in A} \hat{g}_{ij} y_{ij} + \sum_{(i,j) \in A} \sum_{p \in P} \hat{c}_{ij} x_{ij}^p$$

subject to

$$\sum_{j \in V_i^+} x_{ij}^p - \sum_{j \in V_i^-} x_{ji}^p = d_i^p \qquad \forall i \in V, p \in P$$

$$x_{ij}^p \geq 0 \qquad \forall (i,j) \in A, p \in P$$

$$y_{ij} \in \{0,1\} \qquad \forall (i,j) \in A,$$

where $\hat{g}_{ij} = g_{ij} - \beta_{ij} u_{ij}$ and $\hat{c}_{ij}^p = c_{ij}^p + \beta_{ij}$ are the *reduced costs* of the relaxed problem. It is easy to see that LR_1 decomposes into two subproblems, where the first subproblem, shown here, is solely in the y variables.

$$(SP_1(\beta)) \quad \text{Minimise} \sum_{(i,j) \in A} \hat{g}_{ij} y_{ij}$$

subject to

$$y_{ij} \in \{0,1\} \quad \forall (i,j) \in A.$$

An optimal solution y^* of this problem can be found by inspection by setting $y_{ij}^* = 1$ if $\hat{g}_{ij} < 0$, and $y_{ij}^* = 0$ otherwise. The second subproblem is defined only over the x variables and is given as follows:

$$(SP_2(\beta)) \quad \text{Minimise} \sum_{(i,j) \in A} \sum_{p \in P} \hat{c}_{ij} x_{ij}^p$$

subject to

$$\sum_{j \in V_i^+} x_{ij}^p - \sum_{j \in V_i^-} x_{ji}^p = d_i^p \quad \forall i \in V, p \in P$$

$$x_{ij}^p \geq 0 \quad \forall (i,j) \in A, p \in P,$$

which further decomposes into $|P|$ single commodity minimum cost network flow problems. Each of these can be cast and solved as a shortest-path problem, where the costs of the arcs are given by $\hat{c}_{ij}$. The following example illustrates the application of Lagrangean relaxation coupled with subgradient optimisation for an instance of the capacitated network design problem.

Example 8.3

Consider an instance of the capacitated network design problem defined on the same network as shown in Figure 8.1 using the same fixed and variable costs of the network. The additional parameters relating to arc capacities are given as $u_{12} = 3, u_{14} = 1, u_{23} = 2, u_{24} = 1$ and $u_{34} = 1$. We will once again assume that there are two commodities, shown by p_1 and p_2, to be sent over the network. In particular, one unit of commodity p_1 originates from node 1 and is destined to node 3 and *two* units of commodity p_2 from origin node 1 to destination node 4.

Due to arc capacities, it is clear that any feasible solution of the instance will require that commodity p_2 be split and routed through different paths. Indeed, the optimal solution (y^*, x^*) of this instance as shown in Figure 8.2 shows the optimal design, where commodity p_1 is routed on the path $(1, 2, 3)$, one unit of p_2 is routed on path $(1, 2, 4)$ and another unit on $(1, 3, 4)$. The optimal value of the problem is equal to 100.

A step-by-step application of Lagrangean relaxation and subgradient optimisation is shown here:

0. We denote the Lagrangean multipliers using the vector

$$\beta = (\beta_{12}, \beta_{14}, \beta_{23}, \beta_{24}, \beta_{34}),$$

and the subgradients using the vector

$$\sigma = (\sigma_{12}, \sigma_{14}, \sigma_{23}, \sigma_{24}, \sigma_{34}).$$

A subgradient σ_{ij} is defined for each arc $(i, j) \in A$ and corresponds to the following:

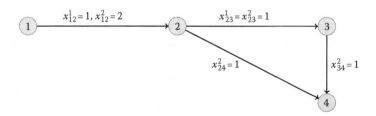

Figure 8.2 Optimal solution of the capacitated network design problem instance.

$$\sum_{p \in P} \bar{x}_{ij}^p - u_{ij}\bar{y}_{ij},$$

where $(\bar{y}, \bar{x})$ is a vector denoting a solution of the relaxed problem $LR_1(\beta)$. The solution vectors are defined as follows:

- $\bar{y} = (\bar{y}_{12}, \bar{y}_{14}, \bar{y}_{23}, \bar{y}_{24}, \bar{y}_{34})$,
- $\bar{x} = (\bar{x}^1, \bar{x}^2)$,
- $\bar{x}^1 = (\bar{x}_{12}^1, \bar{x}_{14}^1, \bar{x}_{23}^1, \bar{x}_{24}^1, \bar{x}_{25}^1)$,
- $\bar{x}^2 = (\bar{x}_{12}^2, \bar{x}_{14}^2, \bar{x}_{23}^2, \bar{x}_{24}^2, \bar{x}_{25}^2)$.

We initially set $\beta^1 = 0$, also known as a *cold start*. It is possible to use a non-trivial set of initial values for the Lagrangean multipliers, known as a *warm start*, if good approximations of the optimal values of the multipliers are available. In the following, we use $\lambda = 1$ in the calculation of the step-length.

An upper bound for the instance can be computed in a fairly naive way, by activating arcs $(1, 2)$, $(1, 4)$, $(2, 3)$ and $(2, 4)$ and setting the flow variables as $x_{12}^1 = x_{24}^1 = x_{12}^2 = x_{23}^2 = 1$, yielding $UB = 105$, which is suboptimal.

1. As $\beta^1 = 0$, we have $\hat{g}_{ij} = g_{ij}$ and $\hat{c}_{ij}^p = c_{ij}^p$, as shown in the following table:

Arcs	(1, 2)	(1, 4)	(2, 3)	(2, 4)	(3, 4)
$\hat{g}_{ij}$	5	10	25	15	20
$\hat{c}_{ij}^1$	5	30	5	5	5
$\hat{c}_{ij}^2$	5	30	5	5	5

Solving $SP_1(\beta^1)$ yields the optimal solution $\bar{y} = 0$. On the other hand, the optimal solution of $SP_2(\beta^1)$ is $\bar{x}^1 = (1, 0, 1, 0, 0)$ and $\bar{x}^2 = (2, 0, 0, 2, 0)$, yielding a $LB = 30$. The subgradient vector corresponding to this solution is $\sigma^1 = (3, 0, 1, 2, 0)$, using which the step-length is calculated as $s^1 = (105 - 30)/(3^2 + 0^2 + 1^2 + 2^2 + 0^2) = 5.36$. The multipliers are updated as $\beta^2 = (16.07, 0, 5.36, 10.71, 0)$.

2. The new Lagrangean multipliers β^2 are used to update the reduced costs as shown in the following table:

Arcs	(1, 2)	(1, 4)	(2, 3)	(2, 4)	(3, 4)
$\hat{g}_{ij}$	−43.21	10	14.29	4.29	20
$\hat{c}_{ij}^1$	21.07	30	10.36	165.71	5
$\hat{c}_{ij}^2$	21.07	30	10.36	165.71	5

Using these new costs, the optimal solution of $SP_1(\beta^2)$ is $\bar{y} = (1,0,0,0,0)$ and the optimal solution of $SP_2(\beta^2)$ is $\bar{x}^1 = (1,0,1,0,0)$ and $\bar{x}^1 = (0,2,0,0,0)$, yielding a $LB = 48.21$. The subgradient vector corresponding to this solution is $\sigma^2 = (-2,2,1,0,0)$, using which the step-length is calculated as $s^2 = (105 - 48.21)/(-2^2+2^2+1^2+0^2+0^2) = 6.31$. The multipliers are updated as $\beta^3 = (3.45, 12.62, 11.67, 10.71, 0)$.

3. In this iteration, the reduced costs for the subproblems are computed as shown in the following table:

Arcs	(1, 2)	(1, 4)	(2, 3)	(2, 4)	(3, 4)
$\hat{g}_{ij}$	−5.35	−2.62	1.67	4.29	20
$\hat{c}_{ij}^1$	8.45	42.62	16.67	15.71	5
$\hat{c}_{ij}^2$	8.45	42.62	16.67	15.71	5

With the new set of costs, the optimal solutions of the subproblems $SP_1(\beta^3)$ and $SP_2(\beta^3)$ are $\bar{y} = (1,1,0,0,0)$, $\bar{x}^1 = (1,0,1,0,0)$ and $\bar{x}^1 = (2,0,2,0,0)$, which jointly imply a $LB = 65.47$. The subgradient vector corresponding to this solution is $\sigma^3 = (-2,-1,1,2,0)$, using which the step-length is calculated as $s^1 = (105 - 65.47)/(-2^2 + -1^2 + 1^2 + 2^2 + 0^2) = 5.68$. The multipliers are updated as $(-7.91, 6.94, 17.35, 22.07, 0)$. However, as the Lagrangean multipliers β should be nonnegative, the multiplier vector is updated as $\beta^4 = (0, 6.94, 17.35, 22.07, 0)$, where $\beta_{12}^4 = \max\{0, -7.91\} = 0$.

4. A new set of multipliers are used to compute the new values of the reduced costs as shown in the following table:

Arcs	(1, 2)	(1, 4)	(2, 3)	(2, 4)	(3, 4)
$\hat{g}_{ij}$	5	3.06	−9.69	−7.07	20
$\hat{c}_{ij}^1$	5	36.94	22.35	27.07	5
$\hat{c}_{ij}^2$	5	36.94	22.35	27.07	5

With the new set of costs, the optimal solutions of the subproblems $SP_1(\beta^4)$ and $SP_2(\beta^4)$ are $\bar{y} = (0,0,1,1,0)$, $\bar{x}^1 = (1,0,1,0,0)$ and $\bar{x}^1 = (2,0,2,0,0)$, which jointly imply a $LB = 74.73$.

The reader will observe that there is a monotonic increase in the lower bounds obtained thus far, although subgradient optimisation does not guarantee that this will always be the case. One can continue in the manner described earlier until a desired convergence is achieved or the algorithm reaches a time or an iteration limit.

8.2.4.2 Relaxing the flow constraints

An alternative relaxation can be obtained when the flow constraints (8.9) are relaxed using unrestricted multipliers α_i^p, in which case the relaxed problem then takes the following form:

$$(LR_2(\alpha)) \quad \text{Minimise} \sum_{(i,j)\in A} g_{ij}y_{ij} + \sum_{(i,j)\in A}\sum_{p\in P} \bar{c}_{ij}^p x_{ij}^p + \sum_{(i,j)\in A}\sum_{p\in P} \alpha_i^p d_i^p$$

subject to

$$\sum_{p\in P} x_{ij}^p \le u_{ij}y_{ij} \quad \forall (i,j) \in A$$

$$x_{ij}^p \ge 0 \qquad \forall (i,j) \in A, p \in P$$

$$y_{ij} \in \{0,1\} \quad \forall (i,j) \in A,$$

where $\bar{c}_{ij}^p = c_{ij}^p + \alpha_j^p - \alpha_i^p$. For a given set of multipliers α_i^p, the last term on the objective is a constant. In this case, problem $LR_2(\alpha)$ decomposes into $|A|$ problems, one for each arc $(i,j) \in A$, each being in the following form:

$$(SP_{ij}(\alpha)) \quad \text{Minimise} \quad g_{ij}y_{ij} + \sum_{p\in P} \bar{c}_{ij}^p x_{ij}^p$$

subject to

$$\sum_{p\in P} x_{ij}^p \le u_{ij}y_{ij}$$

$$x_{ij}^p \ge 0 \qquad \forall p \in P$$

$$y_{ij} \in \{0,1\}.$$

$SP_{ij}(\alpha)$ is a mixed integer programming problem with a *single* binary variable y_{ij}. When $y_{ij} = 1$, let $SP_{ij}^y(\alpha)$ denote the resulting problem, let (x^*,y^*) be an optimal solution and let $v(SP_{ij}^y(\alpha))$ denote the optimal value. Then, it is easy to see that $(x^*,y^*) = (x^*,1)$ if $g_{ij} + v(SP_{ij}^y(\alpha)) < 0$ and $(x^*,y^*) = (0,0)$, otherwise.

The application of subgradient optimisation for the relaxation of the flow constraints can be done in the same way as shown in the previous section.

8.3 Heuristic solution techniques

Today's modern optimisation solvers, such as those listed in Section 8.1, are highly efficient in solving LPs. There also exist efficient algorithms for solving NLPs. When integer variables are involved, such as an MILP or MINLP, the resulting optimisation problems become combinatorial in nature, increasing the difficulty of solving the corresponding formulations to optimality. Even if such formulations can be optimally solved, either with general solvers or bespoke solution techniques such as those presented in Section 8.2, the solution time might make such approaches too slow for use in practice. Although high computational times might not be of much concern for when it comes to solving strategic, or even tactical, problems (such as the facility location problem or the network design problem), it might be of hindrance to operational planners who need to make relatively quick decisions (e.g. such as in the vehicle routing problem), which sometimes may be required in real time (e.g. such as the dynamic vehicle routing problem). If finding an optimal solution for an optimisation problem is computationally costly, this may necessitate the use of methods that are able to provide feasible, but not necessarily optimal, solutions in short computational times.

Heuristics were introduced as methods that are able to search for and produce *near-optimal* solutions in short computational times. A heuristic, in essence, is a search method that starts from one (or a set of) solution(s) and explores a *solution space* by moving from one solution to another. The solution space of the problem is formed by the set of all feasible solutions to the problem, for which an example is given here.

Example 8.4

Consider the TSP instance arising from the parcel delivery problem described in Example 4.1 with one depot and five customers. We provide the distance matrix of the following instance for the sake of completeness (Table 8.1).

There are a total of $5! = 120$ feasible solutions to this instance, each corresponding to a unique ordering of the customers. A graphical representation of the solution space of this instance is shown in

Table 8.1 Distance data (in miles) for the parcel delivery example

	0	1	2	3	4	5
0	0	9.1	12.4	23.3	10.3	19
1	8.8	0	13.3	16.5	10.8	20.3
2	12.8	11.8	0	25.4	23.1	7.5
3	25.4	17.1	25	0	20.8	26.4
4	9.4	10.1	21.8	19.3	0	28.5
5	19.2	20.3	7.6	26.7	29.5	0

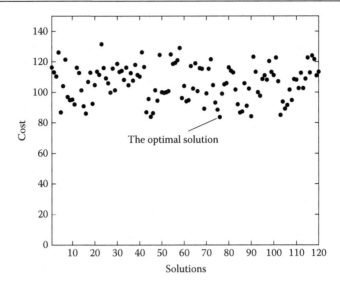

Figure 8.3 A representation of the solution space for the TSP instance of Example 4.1.

Figure 8.3, where the horizontal axis represents the index of the individual solutions of the problem, listed in no particular order. The vertical axis shows the value of each solution, namely the total distance of the corresponding tour. The figure also shows the location of the optimal solution with value 83.7.

An alternative representation of the solution space where the solutions are ordered in a non-decreasing order of their values is given in Figure 8.4, which makes it much easier to locate the optimal solution. With the exception of very small-size instances, however, it is impractical to list all the feasible solutions to be able to identify an optimal solution to an optimisation problem. For this reason, an explicit representation of the solution space of an optimisation problem is generally not available, and one would have to search through it using various mechanisms to guide the search.

The one drawback of heuristic algorithms is that they are unable to provide a guarantee of optimality. In other words, when a heuristic algorithm identifies a feasible solution s to an optimisation problem, it is unable to confirm whether s is optimal or provide any information as to the quality of s. The latter is generally measured as the percentage deviation of the value $v(s)$ of solution s from the optimal value v^* of the problem, if known. For a problem of a minimisation type, the value $v(s)$ is also known as an *upper bound*, using which the percentage deviation is calculated as follows:

$$100 \times \frac{v(s) - v^*}{v^*}.$$

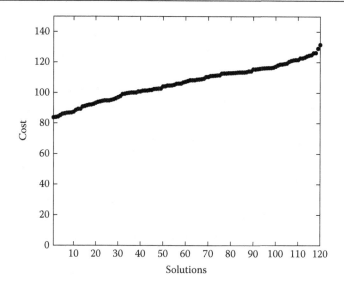

Figure 8.4 An ordered representation of solutions for the TSP instance of Example 4.1.

If v^* is not known, then a *lower bound* v_L for the problem may be used instead.

There are three main mechanisms that form the building blocks of a heuristic algorithm, each of which addresses, respectively, the three questions listed here:

1. *Where in the solution space does one start the search?* This question relates to identifying an initial solution from where the exploration of the solution space will begin. This requires the use of a problem-specific procedure, also known as a *constructive heuristic*.
2. *How does one identify the next solution to move to?* This question is relevant to the concept of a *neighbourhood* of a solution, defined and generated with respect to an *operator*, each containing the *neighbours* of that solution. There may be a number of operators applicable to a given solution, each of which gives rise to a unique neighbourhood.
3. *How does one decide whether to move to another solution or not?* Once a neighbourhood of a solution is generated, and a neighbouring solution is identified, this question seeks to find out whether one should move to that solution or not. This can be done through the use of an *evaluation function*.

The three mechanisms just introduced will be described in more detail here.

8.3.1 Constructive heuristics

A constructive heuristic for an optimisation problem is used to identify a feasible, but not necessarily a near-optimal, solution of that problem. The two important features that any constructive heuristic should have are as follows: (1) it should be able to identify a feasible solution by taking all problem-specific constraints into account, and (2) it should be quick in constructing such a solution. The following example shows the application of a very simple constructive heuristic that can be used to obtain a feasible solution to the travelling salesman problem.

Example 8.5

Consider, once again, the TSP instance described in Example 4.1. To identify a feasible solution, we will use a procedure named the *nearest neighbour heuristic*. This procedure starts from the depot, assigns it as the current node i, then selects a customer j with the minimum cost c_{ij} over all customers not yet visited, sets j as the current node i and iterates in this manner until all nodes are visited. The heuristic follows these steps:

1. We first start from the depot by selecting $i = 0$. The first customer to visit on the route is chosen by the following:

$$\underset{\{1,...,5\}}{\operatorname{argmin}}\{c_{01}, c_{02}, c_{03}, c_{04}, c_{05}\} = \{9.1, 12.4, 23.3, 10.3, 19\},$$

which returns node 1 as the closest customer. The partial route constructed so far is given by $(0, 1)$.

2. Let $i = 1$, and the next customer to visit is given by

$$\underset{\{2,...,5\}}{\operatorname{argmin}}\{c_{12}, c_{13}, c_{14}, c_{15}\} = \{13.3, 16.5, 10.8, 20.3\} = 4.$$

The updated route is $(0, 1, 4)$.

3. Let $i = 4$. Of the three remaining customers $\{2, 3, 5\}$, the one with the shortest distance from node 4 is node 3, using which the partial route is updated as $(0, 1, 4, 3)$.

4. From $i = 3$, the closest of the two customers $\{2, 5\}$ not yet visited is node 2, which implies that node 5 will be the last customer visited before returning to the depot, resulting in the final (feasible) route $(0, 1, 4, 3, 2, 5, 0)$.

The cost of the initial route $(0, 1, 4, 3, 2, 5)$ is 90.90, which implies a $100 \times (90.90 - 83.7)/83.7 = 8.6\%$ deviation from the optimal value.

We present another example for constructive heuristic, this time for the capacitated vehicle routing problem, which is the popular method of Clarke and Wright (1964).

Table 8.2 Savings $\hat{c}_{ij}$ for the distance data of Example 4.7

	1	2	3	4	5
1	–	7.9	15.6	8.3	7.5
2	10.1	–	10.7	0	24.3
3	17.4	12.8	–	14.9	18
4	8.4	0	13.4	0	−0.1
5	8	24	15.8	0	–

Example 8.6

For a vehicle routing problem with n nodes and a common vehicle capacity π, the Clarke and Wright (1964) heuristic starts by constructing n tours of the form $(0, i, 0)$, one for each customer $i = 1, \ldots, n$, and seeks to merge pairs of tours to reduce the overall cost of the solution as much as possible without violating the constraints. In other words, the two routes $(0, \ldots, i, 0)$ and $(0, j, \ldots, 0)$ can be merged into the route $(0, \ldots, i, j, \ldots, 0)$ if the savings $\hat{c}_{ij} = c_{i0} + c_{0j} - c_{ij}$ for all $i, j \in V_C$, $i \neq j$ are positive and the total demand on the combined route does not exceed π. The procedure iteratively merges routes, provided that the merge does not destroy previously selected arcs.

The savings computed for the distance data of Example 4.7 are shown in Table 8.2. The application of the heuristic proceeds with constructing five routes, namely $(0, 1, 0)$, $(0, 2, 0)$, $(0, 3, 0)$, $(0, 4, 0)$ and $(0, 5, 0)$, where $q_1 = 2$, $q_2 = 1$, $q_3 = 2$, $q_4 = 3$, $q_5 = 3$ and $\pi = 4$.

To identify the routes to merge, it suffices to look at all arcs (i, j) formed between the customers for which the savings $\hat{c}_{ij}$ are positive. While any such arc can be chosen for merging routes, a commonly used rule of thumb is to select arcs in a non-increasing order of the savings. The largest saving for this instance is given by the arc $(2, 5)$, for which $\hat{c}_{25} = 24.3$. For this reason, the first iteration of the algorithm merges routes $(0, 2, 0)$ and $(0, 5, 0)$ into $(0, 2, 5, 0)$, which is feasible as it does not violate the capacity constraint. The next largest saving in the matrix is given by $\hat{c}_{52} = 24.3$, but as this arc is already used, the procedure moves on to arc $(3, 5)$ that provides the third largest saving, namely $\hat{c}_{35} = 18$. However, this arc cannot be used for a merge, not only because it would mean breaking arc $(2, 5)$ but also because it violates the capacity constraint due to $q_3 + q_5 = 5 > \pi$. Continuing in a similar fashion, the routes $(0, 3, 0)$ and $(0, 1, 0)$ are merged into $(0, 3, 1, 0)$ given $\hat{c}_{31} = 17.4$, which saturates the total capacity on this route. As further merges of routes are not possible, the procedure stops.

The application of the Clarke and Wright (1964) heuristic on the instance of Example 4.7 has identified a solution that is also optimal. However, this is not always the case and the heuristic might well yield an initial feasible solution with a cost that is significantly higher than that of the optimal.

We now explore how neighbours of a solution can be generated.

8.3.2 Neighbourhoods

A neighbour s' of a (feasible) solution s is one that is obtained by applying a particular operator o to s that results in a different solution. The neighbourhood $\mathcal{N}_o(s)$ of s is obtained by all the neighbours obtained by applying operator o to solution s.

Operators are often problem-specific. In the following example, we show how a particular operator can be applied to the travelling salesman problem.

> **Example 8.7**
>
> We describe a simple operator that takes, as input, a feasible solution s to a TSP represented by a sequence $s = (v_0, v_1, v_2, \ldots, v_n, v_{n+1})$ of nodes, where $v_{n+1} = v_0$. An *insert* operator $\mathcal{I}$ then selects a node v_i^*, $i = 1, \ldots, n$, and inserts the node between all possible pairs of nodes (v_k, v_{k+1}) where $k = 0, \ldots, n$. The insert neighbourhood $N_{\mathcal{I}(s)}$ of solution s then consists of all the solutions obtained by applying operator $\mathcal{I}$ to all nodes v_i $(i = 1, \ldots, n)$ in the solution.
>
> Consider now the solution constructed in Example 8.5 for the delivery example, namely $(0, 1, 4, 3, 2, 5, 0)$. Applying the insert operator on this solution by selecting $v_i^* = 1$ gives rise to the four following solutions: solution $(0, 4, 1, 3, 2, 5, 0)$ with cost 88.60, solution $(0, 4, 3, 1, 2, 5, 0)$ with cost 86.7, solution $(0, 4, 3, 2, 1, 5, 0)$ with cost 105.9 and solution $(0, 4, 3, 2, 5, 1, 0)$ with cost 91.20.

A neighbour solution $s' \in \mathcal{N}_o(s)$ need not necessarily be feasible during the search. Here, we provide an example to illustrate how infeasible solutions may be encountered in a search if the operators do not take problem-specific features into consideration.

> **Example 8.8**
>
> The insert operator described earlier can be applied to a feasible solution of the vehicle routing problem in two ways: (1) individually on each route of the solution in exactly the same way as in the travelling salesman problem, or (2) between two routes in that a node v_i^* chosen from route r_a is inserted into a position on route r_b.
>
> Consider the instance of Example 8.6 and assume a starting solution with the following four routes: $r_1 = (0, 1, 2, 0)$, $r_2 = (0, 3, 0)$, $r_3 = (0, 4, 0)$ and $r_4 = (0, 5, 0)$, which are all feasible with respect to the vehicle capacity constraint. Table 8.3 shows the application of the insertion operator, where the nodes in the first route r_1, are inserted into the other routes r_2, r_3 and r_4, and indicates whether the resulting routes are feasible with respect to capacity.

Encountering an infeasible solution within a heuristic algorithm does not necessarily indicate the need to discard that solution. Infeasible solutions

Table 8.3 Route feasibility after applying the insertion operator

v_i^*	Insert into route	Feasibility
1	r_2	Yes
2	r_2	Yes
1	r_3	No
2	r_3	Yes
1	r_4	No
2	r_4	Yes

can *temporarily* be allowed within the search, subject to the heuristic algorithm ultimately terminating with a feasible solution. This can be achieved by either (1) restoring the infeasibility each time an infeasible solution is identified or (2) by the use of a penalty function that is incorporated in the evaluation of the infeasible solution. The former approach requires new operators that take, as input, an infeasible solution and convert it into a feasible one. The latter is explained further in the next section.

8.3.3 Evaluation function

The quality $v(s')$ of any feasible neighbour solution s' obtained from a given solution s is equal to often the objective function $f(s')$ of that solution. However, if the operator produces an infeasible solution $\hat{s}$, then the evaluation function can be augmented to include a penalty term for any violation of the set of the constraints $g(x) \leq b$ in the form $v(s') = f(s') + \Delta \max\{0, g(x) - b\}$, where Δ is the penalty cost for each unit of violation of the constraints, assuming that the objective function is one of minimisation.

8.3.4 Local search

Local search is a mechanism that starts from a given solution s and explores the neighbourhood of $\mathcal{N}_o(s)$ using a particular operator o, moves to an improved solution s' and repeats the process by starting from s'. The procedure continues until a solution s^* is reached from where no improving solutions in the neighbourhood $\mathcal{N}_o(s^*)$ are found. In this case, s^* is said to be a *locally optimal point*, or simply a local optimum. Sometimes, a time limit can instead be imposed on the algorithm.

The pseudocode of a generic local search algorithm for a minimisation problem is given in Algorithm 8.3.

In Algorithm 8.3, the first two lines define the input parameters, including an initial starting solution s, an operator o and an indicator parameter LocalOptimum that is used to determine whether the algorithm has identified a local optimum or not. The algorithm searches for all the neighbours of solution s within the set $\mathcal{N}_o(s)$ in line 5 and identifies if there are any neighbouring solutions that yield an improvement in the evaluation

---

Algorithm 8.3 Local search

1: **Input** solution s and an operator o.
2: LocalOptimum $= 0$.
3: Set $\hat{s} = s$.
4: **while** LocalOptimum $= 0$ **do**
5:     **for all** $s' \in \mathcal{N}_o(s)$ **do**
6:         **if** $v(s') < v(\hat{s})$ **then**
7:             Set $\hat{s} = s'$
8:         **else**
9:             LocalOptimum $= 1$.
10:        **end if**
11:    **end for**
12:    Set $s = \hat{s}$.
13: **end while**
14: s is a local optimum.

---

function $v(s)$. If there exists at least one improving solution, lines 6 and 7 identify the solution $\hat{s}$ that yields the maximum improvement, where ties are broken arbitrarily. The new solution is replaced by the current solution s. This process is repeated anytime a new solution is found. Note that some solutions are repeated in a given iteration as the insertion operator applied on a tour using two different nodes can result in the same tour.

An application of the local search procedure is presented in the following example.

Example 8.9

We apply Algorithm 8.3 to the instance described in Example 8.5 using the initial solution $s = (0, 1, 4, 3, 2, 5, 0)$ constructed by the nearest neighbour heuristic. The steps are given in Tables 8.4 to 8.6, where the insertion operator is applied to the initial solution in the first iteration, and to best solutions identified in subsequent iterations. The one with the minimum value of $v(s')$ in a given iteration, shown in boldface letters, is chosen as basis for the subsequent iteration. As the tables show, the algorithm terminates in three iterations, where the last iteration fails to identify a solution that improves on the best one found in the second iteration. Consequently, the solution $(0, 4, 1, 3, 5, 2, 0)$ is a local optimum, which, at the same time, is the optimal solution of the original problem for this particular example.

There are two main limitations of local search:

1. Any local search procedure will be *trapped* in a local optimum very quickly. For this reason, it needs to be complemented with additional

Table 8.4 Iterations of the local search algorithm for Example 8.9

Iteration	Initial solution s	Neighbours $s' \in \mathcal{N}_0(s)$	Evaluation function $v(s')$
I	(0, 1, 4, 3, 2, 5, 0)	(0, 4, 1, 3, 2, 5, 0)	88.60
		(0, 4, 3, 1, 2, 5, 0)	86.70
		(0, 4, 3, 2, 1, 5, 0)	105.90
		(0, 4, 3, 2, 5, 1, 0)	91.20
		(0, 4, 1, 3, 2, 5, 0)	88.60
		(0, 1, 3, 4, 2, 5, 0)	94.90
		(0, 1, 3, 2, 4, 5, 0)	121.40
		(0, 1, 3, 2, 5, 4, 0)	97.00
		(0, 3, 1, 4, 2, 5, 0)	99.70
		(0, 1, 3, 4, 2, 5, 0)	94.90
		(0, 1, 4, 2, 3, 5, 0)	112.70
		(0, 1, 4, 2, 5, 3, 0)	101.30
		(0, 2, 1, 4, 3, 5, 0)	99.90
		(0, 1, 2, 4, 3, 5, 0)	110.40
		(0, 1, 4, 2, 3, 5, 0)	112.70
		(0, 1, 4, 3, 5, 2, 0)	86.00
		(0, 5, 1, 4, 3, 2, 0)	107.20
		(0, 1, 5, 4, 3, 2, 0)	116.00
		(0, 1, 4, 5, 3, 2, 0)	112.90
		(0, 1, 4, 3, 5, 2, 0)	**86.00**

Note: Best solution in the neighbourhood shown in bold.

procedures to help *escape* local optima, to be able to explore other parts of the solution space.

2. Using only a single operator within local search may not be sufficient to give the algorithm the ability to effectively explore a solution space. One way to address this limitation is to use multiple operators. In this case, the local search needs a framework within which it will be used, and which will manage the selection and use of the multiple operators.

The consideration of the two points mentioned earlier has given rise to higher-level heuristics, namely *metaheuristics*, which use various mechanisms to overcome these limitations. A number of metaheuristics are presented in the following section.

8.3.5 Metaheuristics

A metaheuristic is a framework which enhances local search with additional mechanisms to improve its effectiveness. A large number and variety of metaheuristics have been proposed in the literature. In what follows, we provide brief descriptions of some of the more popular metaheuristics. A complete description of each metaheuristic is beyond the scope of this

Table 8.5 Iterations of the local search algorithm for Example 8.9

Iteration	Current solution s	Neighbours $s' \in \mathcal{N}_0(s)$	Evaluation function $v(s')$
2	(0, 1, 4, 3, 5, 2, 0)	(0, 4, 1, 3, 5, 2, 0)	83.70
		(0, 4, 3, 1, 5, 2, 0)	87.40
		(0, 4, 3, 5, 1, 2, 0)	102.40
		(0, 4, 3, 5, 2, 1, 0)	84.20
		(0, 4, 1, 3, 5, 2, 0)	**83.70**
		(0, 1, 3, 4, 5, 2, 0)	95.30
		(0, 1, 3, 5, 4, 2, 0)	116.10
		(0, 1, 3, 5, 2, 4, 0)	92.10
		(0, 3, 1, 4, 5, 2, 0)	100.10
		(0, 1, 3, 4, 5, 2, 0)	95.30
		(0, 1, 4, 5, 3, 2, 0)	112.90
		(0, 1, 4, 5, 2, 3, 0)	106.80
		(0, 5, 1, 4, 3, 2, 0)	107.20
		(0, 1, 5, 4, 3, 2, 0)	116.00
		(0, 1, 4, 5, 3, 2, 0)	112.90
		(0, 1, 4, 3, 2, 5, 0)	90.90
		(0, 2, 1, 4, 3, 5, 0)	99.90
		(0, 1, 2, 4, 3, 5, 0)	110.40
		(0, 1, 4, 2, 3, 5, 0)	112.70
		(0, 1, 4, 3, 2, 5, 0)	90.90

Note: Best solution in the neighbourhood shown in bold.

book, but a set of references are provided at the end of the chapter for the interested reader to consult.

8.3.5.1 Simulated annealing

Simulated annealing (SA) is a metaheuristic designed to solve hard combinatorial optimisation problems and mimics the annealing procedure used in metallurgy. Starting with a feasible initial solution, each iteration of the simulated annealing algorithm replaces the current solution by a neighbour solution with a given probability, calculated by taking into account the value of the current solution, the neighbour solution as a possible incumbent and a *temperature* parameter. As is the case in the annealing process, the temperature is progressively decreased throughout the algorithm, and as such, the probability of accepting non-improving solutions decreases as a function of the number of iterations. This mechanism allows the algorithm to perform a relatively high number of uphill moves (i.e. moves that worsen the value of the current solution) during early iterations (providing a means of escaping from a local optimum) but increasingly restricts the probability of such moves as the algorithm progresses.

Table 8.6 Iterations of the local search algorithm for Example 8.9

Iteration	Current solution s	Neighbours s' ∈ $\mathcal{N}_0(s)$	Evaluation function v(s')
3	(0, 4, 1, 3, 5, 2, 0)	(0, 1, 4, 3, 5, 2, 0)	86.00
		(0, 1, 3, 4, 5, 2, 0)	95.30
		(0, 1, 3, 5, 4, 2, 0)	116.10
		(0, 1, 3, 5, 2, 4, 0)	92.10
		(0, 1, 4, 3, 5, 2, 0)	86.00
		(0, 4, 3, 1, 5, 2, 0)	87.40
		(0, 4, 3, 5, 1, 2, 0)	102.40
		(0, 4, 3, 5, 2, 1, 0)	84.20
		(0, 3, 4, 1, 5, 2, 0)	94.90
		(0, 4, 3, 1, 5, 2, 0)	87.40
		(0, 4, 1, 5, 3, 2, 0)	105.20
		(0, 4, 1, 5, 2, 3, 0)	99.10
		(0, 5, 4, 1, 3, 2, 0)	112.90
		(0, 4, 5, 1, 3, 2, 0)	113.40
		(0, 4, 1, 5, 3, 2, 0)	105.20
		(0, 4, 1, 3, 2, 5, 0)	88.60
		(0, 2, 4, 1, 3, 5, 0)	107.70
		(0, 4, 2, 1, 3, 5, 0)	106.00
		(0, 4, 1, 2, 3, 5, 0)	104.70
		(0, 4, 1, 3, 2, 5, 0)	88.60

8.3.5.2 Tabu search

Tabu search (TS) is based on the premise that using a particular move repeatedly in a local search might result in switching between two unique solutions, resulting in what is known as *cycling*. To prevent cycling, TS utilises a *tabu list*, which stores certain attributes of recently visited solutions. Moves within the tabu list are likely to reverse a new solution to an already visited state. Once a move is added to the tabu list, it is forbidden for use for a certain number of iterations, whose value is dictated by the *tabu list size*. By prohibiting the use of such moves, the algorithm prevents the search from selecting moves that lead to previously visited solutions and therefore avoids cycling. TS also uses an *aspiration criterion* which overrides the tabu status of a move in the tabu list, if this move generates a new solution with a value that is better than the best value found so far.

8.3.5.3 Variable neighbourhood descent

The variable neighbourhood descent (VND) algorithm uses a predefined list $(O_1, O_2, \ldots, O_{max})$ of operators, each of which provides a different neighbourhood. The algorithm starts with the first operator O_1 using which it applies local search on a given solution s. When no improvement is possible on s, then the algorithm moves on to the second operator O_2 and iterates in the same way through the list of operators. When an improved solution s' over s is found, the algorithm returns to operator O_1 and applies the list of operators to solution s' in the same iterative manner. The algorithm stops when all operators are exhausted and no improving solutions are identified. For a VND algorithm to be effective, the operators should be such that they should induce dissimilar neighbourhoods.

8.3.5.4 Variable neighbourhood search

Variable neighbourhood search (VNS) is an extension of the VND where, prior to applying local search with a given operator O_i in the set $(O_1, O_2, \ldots, O_m)$, a *shake* mechanism is invoked, which randomly selects a solution in the neighbourhood of O_i. The enumeration through the list of operators and the way that the search restarts are as in VND.

8.3.5.5 Large neighbourhood search

Large neighbourhood search (LNS) is a heuristic that is based on the idea of gradually improving an initial solution by using a destroy operator and a repair operator. The destroy operator removes a part of a given solution. The repair operator inserts the removed components of the solution in a different way so as to obtain a new solution. If the new solution has a better value than the best current solution, LNS replaces the current solution with a new one and uses the new solution as an input to the subsequent iteration. The LNS can be embedded within any metaheuristic, such as SA or TS.

8.3.5.6 Adaptive large neighbourhood search

Adaptive large neighbourhood search (ALNS) is an extension of the LNS heuristic. However, rather than using one large neighbourhood as in LNS, the ALNS applies several removal and insertion operators to a given solution. Each removal and insertion operator i has a certain probability p_i^t of being chosen in every iteration t. The selection criterion is based on the historical performance of every operator and is calibrated by a roulette-wheel mechanism. Initially, the probabilities of each removal and insertion operator are equal. After a certain number of iterations (called a segment), the probability of each operator is recalculated according to its total score. At the start of each segment, the scores of all operators are set to zero. The scores are changed by σ_1 if the best new solution is found, by σ_2 if the new solution is better than the current solution and by σ_3 if the new solution is worse than the current solution. Within the ALNS, a new solution is accepted if it satisfies some criteria defined by the local search framework (e.g. SA) applied at the outer level.

8.3.5.7 Iterated local search

The iterated local search (ILS) algorithm relies on a simple yet effective idea of repeatedly using local search, and perturbing a locally optimum solution when identified at each run of the local search. The perturbation operator is random, and it changes only a part of the locally optimum solution to move the search to other parts of the solution space.

8.3.5.8 Genetic algorithms

All the metaheuristics described so far operate only on a *single* solution s, with the aim of gradually improving the solution by perturbing s or applying local search on s in some form. In other words, these algorithms move from *one* solution to *another* at each step, thereby always maintaining single *incumbent* solution. *Population-based* metaheuristics, on the other hand, use a different philosophy and operate on a *pool* of solutions.

The most popular of population-based metaheuristics is arguably genetic algorithms (GAs). A GA first initialises a pool of solutions, each of which can be generated by using a construction heuristic. GAs employ two main mechanisms, namely *crossover* and *mutation*. The crossover operator takes two feasible solutions s_1 and s_2 as input and, by combining them, obtains two new feasible solutions s_1' and s_2'. Applying the crossover operator repeatedly to pairs of solutions in a given population yields a new population. In order for there to be an improvement from one population to another, GAs use a form of filtering low-quality solutions. Filtering can be done in a number of different ways, one of which is to discard solutions with a poor value. However, being too selective at this step might lead to premature convergence to a pool of solutions, each of which will be a local optimum. To overcome this potential issue, the second mechanism, namely mutation, is applied to a subset of the solutions in a pool. The role of mutation is to perturb a solution, often done in a random fashion. GAs move from one pool of solutions to another by repeatedly using crossover and mutation, until a predefined time or iteration limit is reached, or if sufficiently good-quality solutions are obtained.

8.3.6 Matheuristics

Matheuristics are hybrid methods that combine mathematical programming, or in general optimisation methods, with heuristic algorithms. The aim is to take advantage of the power and capabilities of both classes of methods. Hybridisation can be done in at least two ways, as described here:

1. Embed mathematical programming within a heuristic or a metaheuristic algorithm, in which one or several optimisation problems are solved optimally. This might be the case, for example, when integrated

problems are solved by using a heuristic, but where each individual and smaller problem forming the integrated problem is solved optimally.

2. Use heuristics within the framework of an exact algorithm, in which smaller or subproblems can be solved using a heuristic algorithm. One example to this approach would be to solve the subproblems arising in Benders decomposition or Lagrangean relaxation by using a heuristic or metaheuristic algorithm. A benefit of this approach would be a speed-up in the running time of the algorithm, but this would be at the expense of loss of optimality, or a guarantee thereof. In other words, the bounds provided by such an algorithm would not necessarily be valid.

References and further reading

A wealth of resources exist on exact and heuristic algorithms available to solve mathematical programming formulations. An excellent introduction to linear and integer programming is given by Martin (2012), with a particular focus on large-scale programmes and which presents detailed descriptions of various algorithms, including Benders decomposition, Dantzig-Wolfe decomposition and Lagrangean methods. The reader is referred to Wolsey (1998) and Conforti et al. (2014) for details on the theory of integer programming and to Chen et al. (2010) for a more applied treatment of the topic.

A comprehensive exposition of heuristics and metaheuristics is given in Talbi (2009), which includes single-solution-based metaheuristics, population-based metaheuristics, hybrid metaheuristics, parallel metaheuristics, as well as aspects of implementation, such as parameter tuning and performance analysis. For more details on the individual heuristics mentioned in this chapter, one may refer to the handbooks edited by Glover and Kochenberger (2006) and Gendreau and Potvin (2010). Finally, the paper by Cordeau et al. (2002) provides guidance on various criteria that can be used to select heuristics for the vehicle routing problem, and the survey by Laporte et al. (2014) presents a description of the state-of-the-art heuristic algorithms for the same, but the concepts discussed in these references would also apply to other problems of a similar nature.

Matheuristics are a relatively new area of research; for a survey of applications to routing problems, see Doerner and Schmid (2010) and Archetti and Speranza (2014). The paper by Boschetti and Maniezzo (2009) offers a detailed discussion on integration of metaheuristics with decomposition algorithms, and how the frameworks provided by the latter can be used to guide heuristic search.

References

Achuthan NR and L Caccetta. Integer linear programming formulation for a vehicle routing problem. *European Journal of Operational Research*, 52(1):86–89, 1991.

Adulyasak Y, JF Cordeau and R Jans. Optimization-based adaptive large neighborhood search for the production routing problem. *Transportation Science*, 48(1): 20–45, 2012.

Adulyasak Y, JF Cordeau and R Jans. Formulations and branch-and-cut algorithms for multivehicle production and inventory routing problems. *INFORMS Journal on Computing*, 26(1):103–120, 2013.

Adulyasak Y, JF Cordeau and R Jans. Benders decomposition for production routing under demand uncertainty. *Operations Research*, 63(4):851–867, 2015a.

Adulyasak Y, JF Cordeau and R Jans. The production routing problem: A review of formulations and solution algorithms. *Computers & Operations Research*, 55: 141–152, 2015b.

Agarwal R and Ö Ergun. Ship scheduling and network design for cargo routing in liner shipping. *Transportation Science*, 42(2):175–196, 2008.

Agarwal R, Ö Ergun, L Houghtalen and OO Özener. Collaboration in cargo transportation. In W Chaovalitwongse, KC Furman and PM Pardalos, eds., *Optimization and Logistics Challenges in the Enterprise*, pp. 373–409. Springer, Heidelberg, Germany, 2009.

Albareda-Sambola M. Location-routing and location-arc routing. In G Laporte, S Nickel and F Saldanha da Gama, eds., *Location Science*, pp. 399–418. Springer, Cham, Switzerland, 2015.

Albareda-Sambola M, JA Díaz and E Fernández. A compact model and tight bounds for a combined location-routing problem. *Computers & Operations Research*, 32(3):407–428, 2005.

Alumur S and BY Kara. Network hub location problems: The state of the art. *European Journal of Operational Research*, 190(1):1–21, 2008.

Andersen J and M Christiansen. Designing new European rail freight services. *Journal of the Operational Research Society*, 60(3):348–360, 2009.

Applegate DL, RE Bixby, V Chvatal and WJ Cook. *The Traveling Salesman Problem: A Computational Study*. Princeton University Press, Princeton, NJ, 2011.

Archetti C, L Bertazzi, G Laporte and MG Speranza. A branch-and-cut algorithm for a vendor-managed inventory-routing problem. *Transportation Science*, 41(3): 382–391, 2007.

Archetti C and MG Speranza. A survey on matheuristics for routing problems. *EURO Journal on Computational Optimization*, 2(4):223–246, 2014.

Baldacci R, A Mingozzi and RW Calvo. An exact method for the capacitated location-routing problem. *Operations Research*, 59(5):1284–1296, 2011.

Barnhart C and G Laporte. *Transportation*, Handbooks in Operations Research and Management Science, Vol. 14. Elsevier, Amsterdam, the Netherlands, 2007.

Bauer J, T Bektaş and TG Crainic. Minimizing greenhouse gas emissions in intermodal freight transport: An application to rail service design. *Journal of the Operational Research Society*, 61(3):530–542, 2010.

Bektaş T, M Chouman and TG Crainic. Lagrangean-based decomposition algorithms for multicommodity network design problems with penalized constraints. *Networks*, 55(3):171–180, 2010.

Bektaş T and TG Crainic. A brief overview of intermodal transportation. In D Taylor, ed., *Logistics Engineering Handbook*, pp. 28.1–28.16. CRC Press, Heidelberg, Germany, 2007.

Bektaş T, TG Crainic and V Morency. Improving the performance of rail yards through dynamic reassignments of empty cars. *Transportation Research Part C: Emerging Technologies*, 17(3):259–273, 2009.

Bektaş T and G Laporte. The pollution-routing problem. *Transportation Research Part B: Methodological*, 45(8):1232–1250, 2011.

Bektaş T, PP Repoussis and CD Tarantilis. Dynamic vehicle routing problems. In P Toth and D Vigo, eds., *Vehicle Routing: Problems, Methods and Applications*, Chapter 15, pp. 299–348. SIAM, Philadelphia, PA, 2014.

Belenguer JM, E Benavent, C Prins, C Prodhon and RW Calvo. A branch-and-cut method for the capacitated location-routing problem. *Computers & Operations Research*, 38(6):931–941, 2011.

Benders JF. Partitioning procedures for solving mixed-variables programming problems. *Numerische Mathematik*, 4(1):238–252, 1962.

Bertazzi L and MG Speranza. Inventory routing problems: An introduction. *EURO Journal on Transportation and Logistics*, 1(4):307–326, 2012.

Boschetti M and V Maniezzo. Benders decomposition, Lagrangean relaxation and metaheuristic design. *Journal of Heuristics*, 15(3):283–312, 2009.

Campbell JF. Integer programming formulations of discrete hub location problems. *European Journal of Operational Research*, 72(2):387–405, 1994.

Campbell JF and ME O'Kelly. Twenty-five years of hub location research. *Transportation Science*, 46(2):153–169, 2012.

Chalkiadakis G, E Elkind and M Wooldridge. Computational aspects of cooperative game theory. *Synthesis Lectures on Artificial Intelligence and Machine Learning*, 5(6):1–168, 2012.

Chen DS, RG Batson and Y Dang. *Applied Integer Programming: Modeling and Solution*. Wiley, Hoboken, NJ, 2010.

Christiansen M and K Fagerholt. Ship routing and scheduling in industrial and tramp shipping. In P Toth and D Vigo, eds., *Vehicle Routing: Problems, Methods and Applications*, Chapter 15, pp. 381–408. SIAM, Philadelphia, PA, 2014.

Christiansen M, K Fagerholt, B Nygreen and D Ronen. Maritime transportation. In C Barnhart and G Laporte, eds., *Transportation*, Handbooks in Operations Research and Management Science, Vol. 14, pp. 189–284. Elsevier, Amsterdam, the Netherlands, 2007.

Clarke GU and JW Wright. Scheduling of vehicles from a central depot to a number of delivery points. *Operations Research*, 12(4):568–581, 1964.

Coelho LC, JF Cordeau and G Laporte. Consistency in multi-vehicle inventory-routing. *Transportation Research Part C: Emerging Technologies*, 24:270–287, 2012.

Coelho LC, JF Cordeau and G Laporte. Thirty years of inventory routing. *Transportation Science*, 48(1):1–19, 2013.

Coelho LC and G Laporte. A branch-and-cut algorithm for the multi-product multi-vehicle inventory-routing problem. *International Journal of Production Research*, 51(23–24):7156–7169, 2013a.

Coelho LC and G Laporte. The exact solution of several classes of inventory-routing problems. *Computers & Operations Research*, 40(2):558–565, 2013b.

Coelho LC and G Laporte. Improved solutions for inventory-routing problems through valid inequalities and input ordering. *International Journal of Production Economics*, 155:391–397, 2014.

Conforti M, G Cornuéjols and G Zambelli. *Integer Programming*. Springer, Cham, Switzerland, 2014.

Contardo C, JF Cordeau and B Gendron. A computational comparison of flow formulations for the capacitated location-routing problem. *Discrete Optimization*, 10(4):263–295, 2013.

Contreras I, JF Cordeau and G Laporte. Benders decomposition for large-scale uncapacitated hub location. *Operations Research*, 59(6):1477–1490, 2011a.

Contreras I, JF Cordeau and G Laporte. Exact solution of large-scale hub location problems with multiple capacity levels. *Transportation Science*, 46(4):439–459, 2012.

Contreras I, JA Díaz and E Fernández. Branch and price for large-scale capacitated hub location problems with single assignment. *INFORMS Journal on Computing*, 23(1):41–55, 2011b.

Cook W. *Pursuit of the Traveling Salesman: Mathematics at the Limits of Computation*. Princeton University Press, Princeton, NJ, 2012.

Cordeau JF, M Gendreau, G Laporte, JY Potvin and F Semet. A guide to vehicle routing heuristics. *Journal of the Operational Research Society*, 53(5):512–522, 2002.

Cordeau JF, G Laporte, MWP Savelsbergh and D Vigo. Vehicle routing. In C Barnhart and G Laporte, eds., *Transportation*, Handbooks in Operations Research and Management Science, Vol. 14, pp. 367–428. Elsevier, Amsterdam, the Netherlands, 2007.

Crainic TG. Service network design in freight transportation. *European Journal of Operational Research*, 122(2):272–288, 2000.

Crainic TG. Intermodal transportation. In C Barnhart and G Laporte, eds., *Transportation*, Handbooks in Operations Research and Management Science, Vol. 14, pp. 467–537. Elsevier, Amsterdam, the Netherlands, 2007.

Crainic TG, JA Ferland and JM Rousseau. A tactical planning model for rail freight transportation. *Transportation Science*, 18(2):165–184, 1984.

Crainic TG and JM Rousseau. Multicommodity, multimode freight transportation: A general modeling and algorithmic framework for the service network design problem. *Transportation Research Part B: Methodological*, 20(3):225–242, 1986.

Dantzig G, R Fulkerson and S Johnson. Solution of a large-scale traveling-salesman problem. *Journal of the Operations Research Society of America*, 2(4):393–410, 1954.

Dantzig GB and JH Ramser. The truck dispatching problem. *Management Science*, 6(1):80–91, 1959.

de Camargo RS, G de Miranda Jr, RPM Ferreira and HPL Luna. Multiple allocation hub-and-spoke network design under hub congestion. *Computers & Operations Research*, 36(12):3097–3106, 2009.

Demir E, T Bektaş and G Laporte. A comparative analysis of several vehicle emission models for road freight transportation. *Transportation Research Part D: Transport and Environment*, 16(5):347–357, 2011.

Demir E, T Bektaş and G Laporte. An adaptive large neighborhood search heuristic for the pollution-routing problem. *European Journal of Operational Research*, 223(2):346–359, 2012.

Demir E, T Bektaş and G Laporte. The bi-objective pollution-routing problem. *European Journal of Operational Research*, 232(3):464–478, 2014a.

Demir E, T Bektaş and G Laporte. A review of recent research on green road freight transportation. *European Journal of Operational Research*, 237(3):775–793, 2014b.

Desrochers M and G Laporte. Improvements and extensions to the Miller–Tucker–Zemlin subtour elimination constraints. *Operations Research Letters*, 10(1): 27–36, 1991.

Doerner KF and V Schmid. Survey: Matheuristics for rich vehicle routing problems. In MJ Blesa, C Blum, G Raidl, A Roli and M Sampels, eds., *Hybrid Metaheuristics: Proceedings of the Seventh International Workshop, HM 2010*, Vienna, Austria, 1–2 October 2010, pp. 206–221. Springer, Berlin, Germany, 2010.

Drexl M and M Schneider. A survey of variants and extensions of the location-routing problem. *European Journal of Operational Research*, 241(2):283–308, 2015.

Eglese R and T Bektaş. Green vehicle routing. In P Toth and D Vigo, eds., *Vehicle Routing: Problems, Methods and Applications*, Chapter 15, pp. 437–458. SIAM, Philadelphia, PA, 2014.

Ehmke JF, M Campbell and BW Thomas. Vehicle routing to minimize time-dependent emissions in urban areas. *European Journal of Operational Research*, 251(2):478–494, 2016.

Elhedhli S and FX Hu. Hub-and-spoke network design with congestion. *Computers & Operations Research*, 32(6):1615–1632, 2005.

Erdoğan G, F McLeod, T Cherrett and T Bektaş. Matheuristics for solving a multi-attribute collection problem for a charity organisation. *Journal of the Operational Research Society*, 66(2):177–190, 2015.

Evers PT, DV Harper and PM Needham. The determinants of shipper perceptions of modes. *Transportation Journal*, 36(2):13–25, 1996.

Evers PT and CJ Johnson. Performance perceptions, satisfaction, and intention: The intermodal shipper's perspective. *Transportation Journal*, 40(2):27–39, 2000.

Fagerholt K. Ship scheduling with soft time windows: An optimisation based approach. *European Journal of Operational Research*, 131(3):559–571, 2001.

Fagerholt K, G Laporte and I Norstad. Reducing fuel emissions by optimizing speed on shipping routes. *Journal of the Operational Research Society*, 61(3):523–529, 2010.

Farahani RZ, M Hekmatfar, AB Arabani and E Nikbakhsh. Hub location problems: A review of models, classification, solution techniques, and applications. *Computers & Industrial Engineering*, 64(4):1096–1109, 2013.

Franceschetti A, D Honhon, T Van Woensel, T Bektaş and G Laporte. The time-dependent pollution-routing problem. *Transportation Research Part B: Methodological*, 56:265–293, 2013.

Frisk M, M Göthe-Lundgren, K Jörnsten and M Rönnqvist. Cost allocation in collaborative forest transportation. *European Journal of Operational Research*, 205 (2):448–458, 2010.

Gavish B and SC Graves. The travelling salesman problem and related problems. Technical report, Massachusetts Institute of Technology, Operations Research Center, Cambridge, MA, 1978.

Gendreau M, O Jabali and W Rei. Stochastic vehicle routing problems. In P Toth and D Vigo, eds., *Vehicle Routing: Problems, Methods and Applications*, Chapter 15, pp. 213–239. SIAM, Philadelphia, PA, 2014.

Gendreau M and JY Potvin. *Handbook of Metaheuristics*. Springer, Heidelberg, Germany, 2010.

Gendron B, G Crainic and A Frangioni. Multicommodity capacitated network design. In B Sansò and P Soriano, eds., *Telecommunications Network Planning*, Chapter 3, pp. 1–19. Springer, Boston, MA, 1999.

Geoffrion AM. Generalized Benders decomposition. *Journal of Optimization Theory and Applications*, 10(4):237–260, 1972.

Ghiani G, F Guerriero, G Laporte and R Musmanno. Real-time vehicle routing: Solution concepts, algorithms and parallel computing strategies. *European Journal of Operational Research*, 151(1):1–11, 2003.

Ghiani G, G Laporte and R Musmanno. *Introduction to Logistics Systems Planning and Control*. Wiley, Chichester, UK, 2013.

Glover FW and GA Kochenberger. *Handbook of Metaheuristics*. Springer, Heidelberg, Germany, 2006.

Goel A and T Vidal. Hours of service regulations in road freight transport: An optimization-based international assessment. *Transportation Science*, 48(3): 391–412, 2013.

Golden BL, S Raghavan and EA Wasil. *The Vehicle Routing Problem: Latest Advances and New Challenges*, Vol. 43. Springer, New York, 2008.

Gouveia L and JM Pires. The asymmetric travelling salesman problem and a reformulation of the Miller–Tucker–Zemlin constraints. *European Journal of Operational Research*, 112(1):134–146, 1999.

Gouveia L and JM Pires. The asymmetric travelling salesman problem: On generalizations of disaggregated Miller–Tucker–Zemlin constraints. *Discrete Applied Mathematics*, 112(1):129–145, 2001.

Gutin G and AP Punnen. *The Traveling Salesman Problem and Its Variations*. Springer, New York, 2006.

Hvattum LM, I Norstad, K Fagerholt and G Laporte. Analysis of an exact algorithm for the vessel speed optimization problem. *Networks*, 62(2):132–135, 2013.

Kallehauge B. Formulations and exact algorithms for the vehicle routing problem with time windows. *Computers & Operations Research*, 35(7):2307–2330, 2008.

Koç C, T Bektaş, O Jabali and G Laporte. The fleet size and mix pollution-routing problem. *Transportation Research Part B: Methodological*, 70:239–254, 2014.

Koç C, T Bektaş, O Jabali and G Laporte. The impact of depot location, fleet composition and routing on emissions in city logistics. *Transportation Research Part B: Methodological*, 84:81–102, 2016a.

Koç C, T Bektaş, O Jabali and G Laporte. Thirty years of heterogeneous vehicle routing. *European Journal of Operational Research*, 249(1):1–21, 2016b.

Kontovas CA and HN Psaraftis. Transportation emissions: Some basics. In HN Psaraftis, ed., *Green Transportation Logistics*, Chapter 2, pp. 41–79. Springer, Heidelberg, Germany, 2016.

Krajewska MA and H Kopfer. Collaborating freight forwarding enterprises. *OR Spectrum*, 28(3):301–317, 2006.

Krajewska MA, H Kopfer, G Laporte, S Ropke and G Zaccour. Horizontal cooperation among freight carriers: Request allocation and profit sharing. *Journal of the Operational Research Society*, 59(11):1483–1491, 2008.

Kramer R, N Maculan, A Subramanian and T Vidal. A speed and departure time optimization algorithm for the pollution-routing problem. *European Journal of Operational Research*, 247(3):782–787, 2015a.

Kramer R, A Subramanian, T Vidal and L dos Anjos Cabral. A matheuristic approach for the pollution-routing problem. *European Journal of Operational Research*, 243(2):523–539, 2015b.

Kulkarni RV and PR Bhave. Integer programming formulations of vehicle routing problems. *European Journal of Operational Research*, 20(1):58–67, 1985.

Laporte G. The vehicle routing problem: An overview of exact and approximate algorithms. *European Journal of Operational Research*, 59(3):345–358, 1992.

Laporte G. Fifty years of vehicle routing. *Transportation Science*, 43(4):408–416, 2009.

Laporte G, S Nickel and F Saldanha da Gama. *Location Science*. Springer, Cham, Switzerland, 2015.

Laporte G, Y Nobert and D Arpin. An exact algorithm for solving a capacitated location-routing problem. *Annals of Operations Research*, 6(9):291–310, 1986.

Laporte G, Y Nobert and S Taillefer. Solving a family of multi-depot vehicle routing and location-routing problems. *Transportation Science*, 22(3):161–172, 1988.

Laporte G, S Ropke and T Vidal. Heuristics for the vehicle routing problem. In P Toth and D Vigo, eds., *Vehicle Routing: Problems, Methods and Applications*, Chapter 15, pp. 87–116. SIAM, Philadelphia, PA, 2014.

Lawler EL, K Lenstra, AHG Rinnooy Kan and DB Shmoys. *The Traveling Salesman Problem: A Guided Tour of Combinatorial Optimization*. Wiley, Chichester, UK, 1985.

Magnanti TL and RT Wong. Network design and transportation planning: Models and algorithms. *Transportation Science*, 18(1):1–55, 1984.

Marín A, L Canovas and M Landete. New formulations for the uncapacitated multiple allocation hub location problem. *European Journal of Operational Research*, 172(1):274–292, 2006.

Martello S and P Toth. *Knapsack Problems: Algorithms and Computer Implementations*. Wiley, Chichester, UK, 1990.

Martin RK. *Large Scale Linear and Integer Optimization: A Unified Approach*. Springer, New York, 2012.

McGinnis MA. The relative importance of cost and service in freight transportation choice: Before and after deregulation. *Transportation Journal*, 30(1):12–19, 1990.

McKinnon A. Increasing fuel efficiency in the road freight sector. In A McKinnon, M Browne, M Piecyk and A Whiteing, eds., *Green Logistics: Improving the Environmental Sustainability of Logistics*, Chapter 3, pp. 49–67. Kogan Page, London, UK, 2010.

McKinnon A, M Browne, A Whiteing and M Piecyk. *Green Logistics: Improving the Environmental Sustainability of Logistics*. Kogan Page Publishers, London, UK, 2015.

Miller CE, AW Tucker and RA Zemlin. Integer programming formulation of traveling salesman problems. *Journal of the ACM*, 7(4):326–329, 1960.

Murphy PR and PK Hall. The relative importance of cost and service in freight transportation choice before and after deregulation: An update. *Transportation Journal*, 35(1):30–38, 1995.

Naddef D. A remark on? Integer linear programming formulation for a vehicle routing problem? by NR Achutan and L. Caccetta, or how to use the clark & wright savings to write such integer linear programming formulations. *European Journal of Operational Research*, 75(1):238–241, 1994.

Nagy G and S Salhi. Location-routing: Issues, models and methods. *European Journal of Operational Research*, 177(2):649–672, 2007.

O'Kelly ME. A quadratic integer program for the location of interacting hub facilities. *European Journal of Operational Research*, 32(3):393–404, 1987.

Özener OO and Ö Ergun. Allocating costs in a collaborative transportation procurement network. *Transportation Science*, 42(2):146–165, 2008.

Paraskevopoulos DC, S Gürel and T Bektaş. The congested multicommodity network design problem. *Transportation Research Part E: Logistics and Transportation Review*, 85:166–187, 2016.

Perl J and MS Daskin. A warehouse location-routing problem. *Transportation Research Part B: Methodological*, 19(5):381–396, 1985.

Petersen ER. Railyard modeling: Part II. The effect of yard facilities on congestion. *Transportation Science*, 11(1):50–59, 1977.

Piecyk M. Carbon auditing of companies, supply chains and products. In A McKinnon, M Browne, M Piecyk and A Whiteing, eds., *Green Logistics: Improving the Environmental Sustainability of Logistics*, Chapter 3, pp. 49–67. Kogan Page, London, UK, 2010.

Pillac V, M Gendreau, C Guéret and AL Medaglia. A review of dynamic vehicle routing problems. *European Journal of Operational Research*, 225(1):1–11, 2013.

Powell WB, B Bouzaïene-Ayari and HP Simão. Dynamic models for freight transportation. In C Barnhart and G Laporte, eds., *Transportation*, Handbooks in Operations Research and Management Science, Vol. 14, pp. 285–365. Elsevier, Amsterdam, the Netherlands, 2007.

Prodhon C and C Prins. A survey of recent research on location-routing problems. *European Journal of Operational Research*, 238(1):1–17, 2014.

Psaraftis HN. *Green Transportation Logistics: The Quest for Win-Win Solutions*. Springer, Cham, Switzerland, 2016.

Psaraftis HN and CA Kontovas. Speed models for energy-efficient maritime transportation: A taxonomy and survey. *Transportation Research Part C: Emerging Technologies*, 26:331–351, 2013.

Psaraftis HN and CA Kontovas. Ship speed optimization: Concepts, models and combined speed-routing scenarios. *Transportation Research Part C: Emerging Technologies*, 44:52–69, 2014.

Rushton A, P Croucher and P Baker. *The Handbook of Logistics and Distribution Management: Understanding the Supply Chain*. Kogan Page Publishers, London, UK, 2014.

Simchi-Levi D, P Kaminsky and E Simchi-Levi. *Managing the Supply Chain: The Definitive Guide for the Business Professional*. McGraw-Hill, New York, 2009.

Skorin-Kapov D, J Skorin-Kapov and M O'Kelly. Tight linear programming relaxations of uncapacitated p-hub median problems. *European Journal of Operational Research*, 94(3):582–593, 1996.

Stewart WR and BL Golden. Stochastic vehicle routing: A comprehensive approach. *European Journal of Operational Research*, 14(4):371–385, 1983.

Talbi EG. *Metaheuristics: From Design to Implementation*. Wiley, Hoboken, NJ, 2009.

Tan PZ and BY Kara. A hub covering model for cargo delivery systems. *Networks*, 49(1):28–39, 2007.

Toth P and D Vigo. *Vehicle Routing: Problems, Methods, and Applications*, Vol. 18. SIAM, Philadelphia, PA, 2014.

Verdonck L, A Caris, K Ramaekers and GK Janssens. Collaborative logistics from the perspective of road transportation companies. *Transport Reviews*, 33(6):700–719, 2013.

Waller M, MJ Meixell and M Norbis. A review of the transportation mode choice and carrier selection literature. *The International Journal of Logistics Management*, 19(2):183–211, 2008.

Wang X and H Kopfer. Collaborative transportation planning of less-than-truckload freight. *OR Spectrum*, 36(2):357–380, 2014.

Wolsey LA. *Integer Programming*. Wiley, New York, 1998.

Zhu E, TG Crainic and M Gendreau. Scheduled service network design for freight rail transportation. *Operations Research*, 62(2):383–400, 2014.

Index

Printed in the United States
by Baker & Taylor Publisher Services